RADICAL HOUSING:

ART, XSTRUGGLE, CARE

EDITED BY
ANA VILENICA

Theory on Demand #42

Radical Housing: Art, Struggle, Care

Edited by Ana Vilenica

Cover design: Katja van Stiphout
Design and EPUB development: Tommaso Campagna, Morgane Billuart

Published by the Institute of Network Cultures, Amsterdam, 2021

ISBN PaperBack:
ISBN E-Pub:

Contact

Institute of Network Cultures
Email: info@networkcultures.org
Web: http://www.networkcultures.org

This project has received funding from the European Union's Horizon 2020 research and innovation programme under the Marie Skłodowska-Curie grant agreement No. 707848.

institute of
network cultures

European
Commission

CONTENTS

SECTION IV: NARROWING THE SPACE AND MOURNING THE LOSSES

SECTION V: PATTERNS OF COMMONING AND WORKING WITH DIRT

INTRODUCTION: THE ART OF HOUSING STRUGGLE

ANA VILENICA

But then again, everything has its price in this world

Like this girl's poetry in her step, her lungs

A fair currency, fat with poisonous air. [1]

– Zeena Edwards

It is not enough to acknowledge complicity as an artist or even as a renter. Substantive change requires organizing a movement. [2]

– L.A. Tenants Union

Housing space is a crucial locus of social reproduction, a place where countless acts of care sustain our lives. Yet capital has forced its way into our homes, making them a battleground. For decades now, housing struggles have centered on eviction and foreclosure, along with epidemic homelessness, the segregation of the urban poor, and generalized precarization resulting from worldwide privatization and financialization. These attacks on housing and homes have not gone unanswered. Among practices of resistance honed in the course of the struggle are anti-eviction work, rent strikes, and occupations, whether of land or housing. Struggles have also materialized in informal housing strategies, acts of housing disobedience and collective organizational forms, including tenant unions, activist groups and neighborhood assemblies, drawing on a wide range of political traditions.

It is important to state clearly here that housing struggle is not confined between the walls of any particular home: it constitutes a *housingscape* whose distinct processional and situational character extend its scope and stakes beyond residential accommodation as such.[3] Housing is made, unmade, and remade in this locus of regulation and de-regulation. Within it, housing itself is enacted in various ways by multiple actors contending for dominance and the right to a home. Housingscapes are affirmed and contested, produced, and disrupted in discourses, practices and places, in politics, policy and social struggle, art and culture. All these factors combine to create the paradoxes, contradictions, and instabilities of housing as practice and lived space.

---

1   Zena Edwards, 'A Photo of a Girl – Tribute to the Ogoni 9' in African Writers Abroad and PLATFORM (eds), *No Condition is Permanent, 19 Poets on Climate Justice and Change*, PLATFORM: London, pp. 40– 41.
2   Ultra-red, 'A Response to "Op–Ed: An Ultra-red Line"', X-TRA, 12 Octobar 2017, https://www.x-traonline.org/online/ultra-red-a-response-to-op-ed-an-ultra-red-line [Accessed: 25 May 2021].
3   I developed the *housingscape* concept in this article: Marta Stojić Mitrović and Ana Vilenica, 'Enforcing and Disrupting Circular Movement in an EU Borderscape: Housingscaping in Serbia', *Citizenship Studies* 23, no. 6 (Jun 2019): 540–58.

Art is embedded and intermeshed in housingscapes in multiple ways. The mutation of art into life coincides historically with gradual assimilation of the home – previously experienced as a semi-private, semi-sovereign space for the reproduction of life – into capital and the extractive state. Though art and the question of housing issues have always intersected, the art of housing struggle emerged in contested spaces of the struggle for home. To find an outlet for self-expression, create community spirit and express disagreement, solidarity and call to mobilization, a people's art has depicted what communities in struggle went through. Art makers, whose practices are long-term investment in sight of political and social struggle openly sided against land and housing grab.

Yet, a distinct *art of housing struggle* became known in vacant spaces that appeared in the *transition* from a Fordist to a Post-Fordist model of housing and wider spatial production. Some artists moved into studios, homes and other spaces abandoned in the ruins of industrialization, while others turned their back on crumbling urban environments and settled in rural communes. Artists eventually became significant spatial actors, recognized as such not only within the art world but also in urbanism and real-estate development. This led to a long and complicated history of struggles for artistic and reproductive autonomy, struggles against dispossession and artistic or cultural appropriation whether by the state, elements of civil society or real estate developers.

In recent years, interest has increased in the relationship between art and housing crises. Scholars, policy makers, artists, and activists have begun to address the question in new ways. Meanwhile, many thousands of messages have been exchanged on social media networks debating the issues and contradictions of art in housing struggles. These debates bear witness to the ever-greater difficulty of separating struggles for decent housing (whether one's own or other people's or on a wider collective level) from the question of art production caught in mutations between art institutions and real estate. The stake of art in housing struggles is always someone's home, always potentially one's own. This is precisely what makes it messy and *too connected* to life. This is why the art of housing struggle incessantly generates polemics: one after another, an endless series of passionate calls to choose a side.

Shelter-in-place, self-isolation, stay-at-home and quarantine orders as default worldwide responses to the COVID-19 pandemic have only augmented the existing crisis by bringing housing and health care infrastructure in proximity.[4] Multiple housing crises have affected both the production of art and the reproduction of art producers. As working conditions have become increasingly precarious, access to stable housing, particularly in large cities, has also dwindled, with direct consequences for the artistic practice itself. Artists are expected to draw on their supposed cultural capital to solve the housing crisis, or to live and work by collaborating with the leading (parasitic) agents of regeneration, transition, or colonial modernization. In parallel, artists have formed alliances with oppressed groups, acting as

---

4   Ana Vilenica, Vladimir Mentus, Irena Ristić, 'Struggles for care infrastructures in Serbia: Dispossessed care, coronavirus and housing', Special issue: The pandemic, Dispossessed Care, and Housing, (eds.) Emma Dowling, Birgit Sauer, Ayse Dursun, Verena Kettner und Syntia Hasenoehrl, *Journal for Historical Social Research*, forthcoming in 2021.

militant accomplices in housing struggles,[5] collectively diverting resources to the protection of their homes and those of others.

Just like art, thought itself lays bare a geography. The art of housing struggle has historically been theorized most often from the perspective of the Global North. Knowledge production determines the vocabulary and grammar of the art of housing struggle, which until now has reflected the established Global North/South division.[6] That's why it is important to approach knowledge production from another geographical point of view: from peripheral geographies. How to think the art of housing struggle from perspectives that were never allowed into so-called universalism? My answer to this question would be to adopt a Global East perspective.[7] Since we've always been marginal to debates but influential when it comes to art, I make a case in this introduction for the vantage point of this peripheral geography for the art of housing struggle.[8]

Development-led Art

I participate

You participate

He/She participates

We participate

You participate

They profit.[9]

– Counter-propaganda by Southwark Notes

For many decades now, culture-led urban regeneration has been central to discussion of art and housing struggle(s). Meanwhile, regeneration of designated *declining* urban spaces has

5   Ana Vilenica, 'Becoming an Accomplice in Housing Struggles on Vulturilor Street', *Dialogues in Human Geography* 9, no. 2 (July 2019): 210–13.
6   Ash Amin and Michele Lancione (eds.), *Grammars of the Urban Grounds*, Durham: Duke University Press, forthcoming.
7   Martin Muller, 'Footnote Urbanism: The Missing East in (not so) Global Urbanism in Michele Lancione and Colin McFerlane (eds.), *Global Urbanism: Knowledge, Power and the City*, London: Routledge, 2021, pp. 88-95.
8   My perspective has a foothold in the Global East art – particularly the art from ex-Yugoslav region – as well as in Yugoslav housing experiments that has brought a historically new type of societal property (a property without proprietor) and self-management as a form of a direct democracy in managing housing. Although there were many challenges that were not met in the way it was promised, this history has a potential of a prospective that could – with a proper critical reassessment – bring useful knowledge against capitalist cancelation of the future.
9   The slogan was taken from the poster painted by the French students in the spring of 1968.

become a central focus of urban policy worldwide, with culture promoted as a key means of response to socio-economic decline. In this context, art is expected to produce cultural, social, and economic outcomes, to regenerate the hollowed-out economies of post-industrial cities, and to bind together populations fragmented by capital's systemic reproduction of race, ethnic, religious, national, class, gender, sexual, ableist, geographical divisions – regardless of a total paucity of evidence that the arts can perform any of these tasks. The role of art in housing and urban regeneration reveals its integral involvement in the same ongoing capitalist mutations that reduce the arts to one more capital input.

One significant development in recent critical discourse is a shift in attention from culture-led urban regeneration as such to the identification and criticism of developer-led art, in response to the ever-greater openness and apparent acceptability of the developers' undisguised profit motive. Contributors Josephine Berry and Anthony Iles write: 'as the developers' naked profit principle became an acceptable and open objective, positioning artists increasingly as collateral rather than agents of urban change, and art became a vernacular veneer to be cloned by developers and local authority bureaucrats alike.' The resulting art resembles what Gregory Sholette called *bare art,* or as Kim Charnley describes it in the introduction to Shollette's book *Delirium and Resistance,*[10] a 'new banal obviousness of art's subservience to the interest of capital.' Artists are pushed into this new real estate art market (or markets) by their precarious position of artists and the lack of both public and alternative sources of funding for art production. At the same time, the lack of affordable housing in some parts of the world has made artists more likely to accept various unethical sources of housing and studio space, such as property guardianship on estates undergoing *decanting,* i.e., the clearing-out of existing tenants to free up the space for higher income brackets.[11]

Neither culture-led regeneration nor developer-led art is an entirely coherent concept. Both terms, however, comprehend the *visionary* idea of culture and art as catalyst and engine of optimistic urban change. This vision is embedded in shifting policy driven agendas according to local context.[12] In many instances, markedly similar solutions have been applied to urban issues in the Global East and Global South. In many countries of the Global South and Global East, culture led-urban regeneration and developer-led art were introduced by distinctly neo-colonial means. Worldwide, national public institutions in collaboration with global private developers, supra-national institutions, and multinational NGOs are keen to sponsor socially engaged art projects that promise to perform regenerative tasks.

One example is the *Urban Incubator* project, launched by the Goethe-Institut in Belgrade, Serbia in 2013. The project was initially presented as an art-led participative model of urban development, seeking to improve living conditions in the Savamala neighborhood while democratizing community decision-making processes. Savamala is a neighborhood with

---

10   See Gregory Sholette, *Delirium and Resistance: Activist Art and the Crisis of Capitalism*, London: Pluto Press, 2017.

11   Ana Vilenica, 'The Doomed Pursuit of Dignity: Artists as Property Guardians in and Against Artwashing', *AM Journal of Art and Media Studies,* forthcoming in October 2021.

12   Jonatan Vickery, 'The Emergence of Culture-led Regeneration: A Policy Concept and its Discontent' (Research papers) University of Warwick: Center for Cultural Policy Studies, 2007, p. 6.

specific symbolic importance for post-socialist Serbia. The area is recognized as a site of pre-WWII capitalist development. During the Socialist experiment it became an industrial and infrastructural traffic node, while its pre-war buildings were unmaintained. In the process of capitalist restoration, Savamala became central to the project of re-establishing continuity with the so-called golden age of Serbia.[13]

Fig. 1: Belgrade Waterfront and Savamala. Photo by Nikola Aleksić.[14]

The Goethe-Institut, a German cultural mission, recognized this area as neglected and offered a Western culture-led regeneration model based on resolving spatial deficits with creativization and art. Local and municipal governments supported this project because it was in tune with their goals of legitimizing the history of Savamala and preparing the terrain for future investment. The Goethe-Institut issued an open call, the result of which was that it mostly commissioned Western engaged artists to regenerate the neighborhood. The appointees assumed de facto expert roles as urban planners, social workers, ethnographers, researchers etc., yet few knew anything at all about Savamala.[15] The state used the presence of the artists in its promotion of the formerly declining neighborhood, but it ended the collaboration once the Belgrade Waterfront mega-project

13   Notwithstanding the 90.5% poverty rate in 1929 Belgrade. More about that: Zlata Vuksanović Macura, *Život na ivici: Stanovanje sirotinje u Beogradu 1919-1941*, Beograd: Orion Art, 2012, p. 18.
14   Curtesy to *Unsplash:* https://unsplash.com/s/photos/belgrade-waterfront [Accessed: 21 May 2021].
15   Many of those invited to make art in Savamala had never visited Belgrade before.

was launched. An association originating in the earlier project initially sought to provide 1% of the culture promised by Belgrade Waterfront under the terms of the public-private partnership with the Eagle Hills company of the United Arab Emirates. This soon proved impossible. Critical voices among the artists were silenced in favor of a culture led by developers and the state, among whose outstanding products is a 24-meter national(istic) monument to Stefan Nemanja (*father of the Serbian nation*), the work of an artist from Russia.

In (semi)peripheral regions, the practice of regeneration goes hand in hand with attempts at *modernization* based on Western models, including drastic overhaul of existing socialist and social structures and the introduction of new capitalistic economic forms into urban space, housing, and art world.[16] Cultural processes that foster the subordination of the periphery in the capitalist world-system overlap with a particular developmentalist ideology characterized by a presumption of the inevitability of historical change and by political and economic intervention from the Western core. This ideology became normative within capitalism in the 2000s as a response to the crisis of Post-Keynesian Economy.

The term 'developmentalism' as used here stands both for the ideology of urban growth and progress with no social consciousness, and for that of developmentalism as neo-colonial doctrine pledged to Eurocentric free-market urban development, including the introduction of new financialized and culturalized forms of housing capitalism in the Global South and the Global East. However, developmentalism is not unique to the Global East and the Global South: it also operates within the internal peripheries of the Global North, where the same logic was used to dismantle the *social condensers,*[17] built under the pressure of the well-organized post-WWII working class. Within this process, artists and art take the role of the development agents, called on to perform a particular function which often involves tasks assigned in places wholly unknown to those accepting them.

The artists are called into sites with names like *regeneration sites, modernization projects, transitional landscapes*. Often these are the sites where gentrification is played out, with the urban poor, the working class and black and brown communities displaced to make room for the relatively affluent. 'Development-led art' is a new composite term. I want to propose it here because it connotes both culture-led regeneration *and* developer-led art aspirations, and because it reconnects universalist claims built into these concepts to the ongoing social reality. The term refers to an art practice integrated into neo-colonial real-estate development as a business process encompassing a range of activities such as the refurbishment of existing buildings and neighborhoods, renovation, new building, land transactions and homelessness management.

Development-led art is the art of housing struggle with a function of a tool in efforts of different

---

16   Ana Vilenica, 'The Art of New Class Geography of the City: Culture-Guided Urban Regeneration Serving the Modernization of the Periphery' in Gordana Nikolić and Šefik Tatljić (ed.), *Grey Zones of Creativity and Capital*, Amsterdam: INC, 2014, pp. 51-64.

17   Michał Murawski, 'Introduction: Crystallising the Social Condenser', *The Journal of Architecture 22*, no. 3 (Jun 2017): 372–86. and Michał Murawski and Jane Rendell, 'The Social Condenser: A Century of Revolution Through Architecture, 1917–2017', *The Journal of Architecture 22*, no. 3 (Jun 2017): 369–71.

actors to establish dominion of the exchange value of housing. It has generated various forms of art from obedient bare art to the art of resistance. Critical artists are also a part of this housingscape puzzle as they seek to dignify their position in this unenviable situation by producing social, socially engaged, and critical art, or by disobeying the parameters and rules set by development-led art policy. Some of this art arose from existential angst, some from guilt, while other work emerged from solidarity.

These artists seek to survive generalized crisis, attempting to reflect on exclusionary forms of urban development and at the same time to make a living from the development-led art. Artists have made works denouncing injustices such as local social cleansing; they have contributed to critical archiving and organized performances and expeditions critical of the processes in which they themselves take part. Some artists have sought to empower local communities while participating in the process of their surrender to regeneration through the slow museumification of their life and struggle. In other cases, those open to joining the fight against social cleansing have managed to contribute to it by spreading the word and sharing resources with local communities, thus coming close to the more radical forms of art. Even so, they must always remain cautious about quite how far they will go. The danger is their art might have unwanted personal consequences such as a loss of employment or job studio space or even homelessness.[18] There can be no doubt that the task of preventing developmentalist attacks on local communities is much bigger than the power of art. The pursuit of personal dignity is likewise a limiting goal. Nonetheless, all these experiences should be acknowledged and learned from on the way to joining existing movements against attacks on housing or building new ones.

Radical Housing Art

But the practice of treating housing struggles as a gateway, and the epistemology of dwelling as difference allows us to access the multiple possible forms that radical housing politics can take, for the many, in everyday life.[19]

– Michele Lancione

The art of housing struggle also happens in the (semi)autonomous combative corners, cracks, and grey zones of housingscapes. Since the global housing crisis of 2007–8 and its geographically uneven global effects, the art of housing struggle has continued to engage with radical forms of knowledge production and interventionism around housing struggles and organizing with existing initiatives and housing movements. This can mean taking action against evictions, rent-based violence, etc.; the violence of housing commodification in all its intersectional, class-belligerent, racialized, and gendered dimensions. It implies an art in one-to-one correspondence with the world, 'seeking to escape performative and ontological capture as art altogether,'[20] or using

---

18   Ana Vilenica, 'The Doomed Pursuit of Dignity: Artists as Property Guardians in and Against Artwashing', *AM Journal of Art and Media Studies,* forthcoming in October 2021.
19   Michele Lancione, 'Radical Housing: On the Politics of Dwelling as Difference'. *International Journal of Housing Policy* 20, no. 2 (April 2020): 283.
20   Stephen Wright, *Toward a Lexicon of Usership,* Eindhoven: Van Abbemuseum, 2013, p. 5.

them as a tool in struggle. Some instances of this art practice are consciously undertaken as art; other practitioners reject the art world altogether, while others again develop non-professional art practices.

An *aesthetic of struggle* characteristic of these practices might best be defined in terms of gestures that suggest allyship or complicity[21] (i.e., the condition shared by accomplices) with the communities in struggle, whether or not 'the privilege to co-opt social justice ideologies'[22] is acknowledged. In another sense the aesthetic of struggle also uses an intracommunal lens to express the community's agency amid systemic inequity, creating experience that is not exhausted in oppositionality, knowledge and desire, but instead discovers and generates connections and tensions between the collective and singular, the special and the mundane, as Felix Stalder and Cornelia Sollfrank write in the introduction to their jointly edited volume *Aesthetic of the Commons*. This art lives from a permanent process of experimentation. It envisions and responds to the contemporary moment not only as resistance 'but also as "beautiful" – understanding beauty as that which cannot be fully grasped conceptually nor be simply consumed.'[23]

One clear example of such an art practice is the *Kamendinamics* project by cultural workers Nebojša Milikić and Tadej Kurepa of the Rex Cultural center in Belgrade. It began with a series of conversations with tenants of Kamendin social housing project, located in Zemun polje in Belgrade. The neighborhood attracted public attention after racist protests from neighboring blocks in response to an outbreak of scabies at a local primary school. The protestors accused Roma tenants in Kamendin of spreading the disease and demanded the city official to stop renting flats in the area to Roma. Conversations with tenants soon reviled that the real problem in the neighborhood was debt crises.

State and city officials used the racist unrest to speed up the eviction of hundreds of tenants with outstanding debt resulting from inflated utility bills imposed by the authorities themselves. Answering one survey question, Kamendin tenant Zlatija Kostić declared herself a culprit for her own and her family's predicament, saying she would sue herself. Together with Nebojša Milikić and Tadej Kurepa and few other Kamendin tenants facing eviction, Zlatija Kostić wrote a play in the form of court records from Zlatija's litigation against herself. Zlatija the elder is suing Zlatija the younger over the difficult living conditions and failures that brought her to the council flat in Kamendin despite having worked all her life and under Yugoslavian law having the right to home. The play ends up in a kind of forum theatre situation, where the audience plays the jury. This creates an awkward feeling of being a spectator as well as an accomplice in the situation. Unfortunately, the play has only been performed once in Serbia, but the project itself obtained significant publicity, helping the tenants to build their case against the municipality. The struggle of the Kamendin tenants continues.

21   Ana Vilenica, 'Becoming an Accomplice in Housing Struggles on Vulturilor Street', *Dialogues in Human Geography* 9, no. 2 (July 2019): 210–13.

22   Maya Brooks, 'Art Advocacy: The Aesthetic of Solidarity', *Circa: The NCMA Blog*, 23 Jul 23 2020, https://ncartmuseum.org/blog/view/art_advocacy_the_aesthetic_of_solidarity.

23   Felix Stalder and Cornelia Sollfrank, 'Introduction' in Cornelia Sollfrank, Felix Stalder, and Shusha Niederberger (eds.), *Aesthetic of the Commons*, Zurich: Diaphanes, 2021, p. 32.

Fig. 3 and 4: Collective reading of I Sued my Self at the Art and Housing Struggles conference in London. Photo by Rastko Novaković.

How can we contextualize this specific art practice within the art history of the Global East? What critical terms could we use to understand this specific practice? Could the notion *radical* help in conceptualizing this approach to art in any way? *Radical art* in the interpretation of art in Serbia has been for a long time trapped in artistic avant-gardism. In this approach, radical art is seen as multiplicity of cultural practices stretching from the historic avant-gardes to various postmodern post-avant-gardes, characterized by artistic transgression, experiment and interdisciplinarity:[24] the art practices of the long 20th century that constitute a mainstream in today's art world. Any attempt to reclaim the notion of radicality becomes even more difficult bearing in mind the political use of the word in Serbia. After 1989, the term became synonymous with the extreme nationalism and fanaticism that now rules the country.[25] How can we reclaim the notion of the radical today? Is there a radical art practice in Serbian or Yugoslav art history outside of these genealogies?

Rastko Močnik invokes the radical with reference to 'the symbolic production that was born out of the struggle for life and was perceived by its producers as a weapon in the fight for freedom and emancipation.'[26] The culture construed here as weapon is the Partisan culture and art that emerged in the Anti-Fascist People's Liberation Struggle of Yugoslavia in WWII. Močnik sees Partisan art as an inextricable element of a wider project of political transformation, whereas Miklavž Komelj argues that Partisan art invented a new kind of post-bourgeois artistic autonomy. There is no consensus on the historical origins of this art (its respective relations, for example, to Socialist Realism and the avant-gardes), but there can be no doubt that Partisan art as art of political experiment created a radically new sense of what is historically possible. It is precisely this aspect that makes it essential for the contemporary Art of the Housing Struggle. As Boris Buden wrote in a review of Gal Kirn's book *Partizan Brake* (Partizanski prelomi): 'It is not about past at all, but rather about our own historical present and its radical openness to different futures.'[27]

It is generally agreed that the one-party system of post-war Yugoslavia betrayed the political promise of Partisan art.[28] The conception of radical amateurism nonetheless stands out in regional art history as one instrument worth revisiting within the genealogy of radicality.

The usefulness of this notion, as Aleksandra Sekulić describes it, is that it draws attention to the political, social, and cultural context of Yugoslav cultural and artistic practices rarely mentioned in standard art historiography.[29] Aldo Milohnić writes about this practice as

---

24   Miško Šuvaković, *Istorija umetnosti u Srbiji – XX vek: Radikalne umetničke prakse*, Beograd: Orion Art, 2010, pp. 43–44.
25   In the 1990s this political position was represented by the neo-fascist Serbia Radical Party, led by convicted war criminal Vojislav Šešelj. Current Serbian president Aleksandar Vučić joined the Serbian Radical Party in 1993 and he soon became MP in secretary general of the party.
26   Rastko Močnik, 'The Partisan Symbolic Politics' in *Slavica Tergestina: European Slavic Studies Journal* Vol 17 (2016): 22.
27   Boris Buden, 'It is 'Now', Comrades. On Gal Kirn's Partizanski prelomi' in *Slavica Tergestina: European Slavic Studies Journal* Vol 17 (2016): 167.
28   Gal Kirn, 'The Yugoslav Partisan Art: Introductory Note' in *Slavica Tergestina: European Slavic Studies Journal* Vol 17 (2016): 12.
29   Aleksandra Sekulić, 'Pojam radikalnog amaterizma i mogućnost kontinuiteta posle raspada Jugoslavije',

something opposed to the ideal of autonomous art cherished by the cultural elites of the time. Unburdened by aesthetic concerns, this art turned towards political intervention, towards participation in a 'spontaneous ideology of radical interventionism in cultural, social and political spheres of Yugoslav society.'[30] This artistic and political dilettantism allowed artists-amateurs to ask questions that are not usually asked. Such is the example of the short documentary film from 1971 entitled *Black Film* by Želimir Žilnik. In the middle of the night, the camera follows Žilnik in the town of Novi Sad where he encounters a group of homeless *lumpenproletariat*. Contrary to the official narratives in Yugoslavia that claimed that there is no homelessness, this film sheds light on this issue as Žilnik offers these people to stay in the two-bedroom flat with his family. Is it possible to make a continuity with such practice after Yugoslavia? With the disappearance of Yugoslav cultural infrastructure, radical amateurism became a rare phenomenon. Thus, today we can no longer speak of radical amateurism as an alternative culture, but we *can* speak of radical art practice as form of access, as embodied day-to-day resistance from margins and an element of a movement culture (if such a thing does indeed exist).

Building on Michele Lancione's invitation to expand the radical housing politics by including minor activisms and exploring creative methodologies,[31] I propose the term *radical housing art* for art made by tenants and communities and also for experimental artistic practices that take part in producing radical housing politics against and beyond the development(alist) matrix. Lancione advocates for radical housing 'as a gateway to challenge wider structural forms of violence, including patriarchy, racism, class exploitation, and, of course, deprivation of shelter.'[32] Radical politics is not an exclusive product of Westernized masculine movements with grand narratives, but of mundane acts of subversion made and un-made at the level of everyday dwelling practices and generative power of radical movements that take desperation seriously; desperation as political in itself.[33]

Struggle has been central to the politics of radical housing art, by means of which histories have been reclaimed from below and new infrastructures of care have been built beyond the valorizing structures of art world, accompanied by efforts to facilitate exchange of skills and to raise awareness of multiple possible futures. On this journey, art has been an important tool in thinking about new ways of living and relating to housing and home and mobilizing against housing dispossession. Radical housing art practice also raises questions about whose art counts, and although it has been unable to change the globalized art world, it has managed to create temporary spaces where it is the people's art that counts.

However, radical housing art is not an isolated practice, and the method of *categorical*

Beton No 179, 2017.

30   Aldo Milohnić, 'Radikalni amaterizam', *Priručnik raškolovanog znanja*, Beograd and Skoplje, TkH and Kontrapunkt, 2012, pp. 6.

31   This connection has already been made in: Nitin Bathla and Sumedha Garg, 'Radical housing and socially-engaged art: Reflections from a tenement town in Delhi's extensive urbanisation'. Radical Housing Journal 2, no.2 (2020): 35.

32   Michele Lancione, 'Radical Housing: On the Politics of Dwelling as Difference': 276.

33   Lancione, Radical Housing: 278–279.

isolation,[34] as Marina Vishmidt and Kerstin Stakemeier describe it, must be overcome if we are to understand its potentials and pitfalls. It emerges in a complex network of actors entering housing-related fields from art, the non-profit sector, the state, the private sector, and different social and housing movements. Relations between these various actors have significantly shaped and complicated the meaning, scope, and effects of the radical housing art practices. These art practices and their field are not static or fixed. They have their own dynamics, and it is therefore important to assess them in relation to social, political, economic, and cultural developments on an ongoing basis. They are dynamic and processual, and their constellations change over time and space. If an art practice functions as radical housing art at one moment, this doesn't mean the same will be true when the constellations have changed. That is why it is important to always allow the process of communal becoming to teach us. Radical housing art has its common challenges and common failures, and the role of all of us who fight for homes is to hold one another accountable, as well as to remain aware of the beauty of art that moves from tragic to joyful.

Thinking Radical Housing Collectively

At its core it is about radical humility and making ourselves vulnerable to each other, listening in more ethical ways, overcoming our fear of asking questions, and being OK with not knowing the answers.[35]

– Patricia DeRocher

Though I knew some of them from before, I met most of the contributors to this book at the Art and Housing Struggles: Between Art and Political Organizing conference, which I co-organized with Elena Marchevska and Chris Jones at London South Bank University in June 2018.[36] This conference had two main focuses: global and local. It set out to ask difficult questions about the place of art in the housingscape and its role in ongoing battles worldwide. Besides the session with papers, the conference included two workshops by people whose work was formative to me. Nebojša Milikić and Zlatija Kostić organized a workshop where we all took part in collative reading of their piece, *I Sued My Self.* Dont Rhin, sound artist and popular educator, gave a workshop where he invited participants to reflect on their experiences with gentrification violence. This workshop introduced a framework for transforming experience into an investigation that takes action in the world, based on his work with the School of Echoes Los Angeles that resulted in forming the L. A. Tenants' Union.

London South Bank University is in dangerous proximity to Elephant and Castle Shopping

34   Kerstin Stakemeier, and Marina Vishmidt, *Reproducing Autonomy: Work, Money, Crisis and Contemporary Art*, London: Mute Publishing, 2016.
35   Patricia DeRocher, *Transnational Testimonios: The Politics of Collective Knowledge Production*, University of Washington Press: Seattle, p. XIV.
36   The conference and the research for this article has been conducted within the framework of the European Comision's Horizon 2020 research and innovation programme under the Marie Sklodowska-Curie Individual Fellowship grant for the research project HOUSEREG, agreement No. 707848 (2016–2018).

Centre and Aylesbury Estate, both undergoing social cleansing at the time. That urgent matter and our own involvement with local initiatives against social cleansing in the area made us concentrate on local struggles in London and practices developed by collectives working and organizing with local communities. On the final day of our encounter, Emer Mary Morris and Nina Scott from the You Should See the Other Guy theatre group talked about the practice they developed in housing struggle during the occupation with Focus E15 Mums group. Hannan Majid from the Rainbow Collective and Ledbury Action group talked about the work with children in South East London as part of the latter; a housing campaign created as an emergency response to unsafe and appalling housing conditions on their Council estate. AHHAAHHA from Voices that Shake! talked about their work with youth against social injustice and connections of racial violence, body trauma, and art. Zena Edwards introduced her artistic practice fusing poetry in music to create a soundtrack against racial injustice in the city.

Learning from this encounter, the book provides a composite cartography of modes of thinking and acting around the art of housing struggle. It is a mosaic of interventions that emerge from contradictory traditions, and as a result it proposes heterogeneous orientations. What connects these contributions is thinking from or within the periphery. These peripheries are not located only in the Global South[37] and the Global East but also in the Global North's internal margins. It is in these sights of resistance against housing precarity that radical housing is traced as it emerges, declines, and re-emerges on the way to our common future. The book is divided into five sections. Each section contains essays with a particular focus. The pieces are thought-provoking commentaries on new and old modes of theorizing, thinking, and making art in the midst of housing struggles. Though the authors don't engage directly with the notion of radical housing, this essay collection raises questions about how to produce knowledge and how to do art about and from housing struggles in a radical way.

What is radical politics today, and how to foster radical housing politics at a time of unsettled relations between the accumulation of capital, new forms of social reproduction, new social and housing movements, and the need to rethink the past and envision alternatives? In what way does radical care emerge as a driving force within radical housing movements and resistance at the margins? Can we theorize, criticize, and make art with care or can we care critically? How to write our own insurgent histories radically, and how to keep these histories alive? By what means and with what politics to build tenants' power and housing commons against social cleansing, oppressive forms of housing, urban real estate and artwashing? These are just some of the questions that arise from the essays, between lines and in footnotes, and in ongoing conversations happening as we speak, implicitly or explicitly. The book also raises the issue of knowledge production in Western institutions tackling housing violence and radical housing politics. The argument is made against white colonial institutions and the uncomfortable truth about how these institutions have always treated minoritized people.

---

37   This book suffers from the serious lack of the Global South perspectives and every future attempt to engage with these issues should put an emphasis there. This collection of essays is a product of the circumstances that have emerged during my time in London and a set of decisions that I took due to limited financial capacities of the project and my limited knowledge about other contexts. The research project that I have been working on from 2016 to 2018 had a focus on the UK and Serbia.

The first section, *Insurgent History and Radical Care*, starts with *The Land of Three Towers,* a theatre performance script by the London-based theatre group You Should See the Other Guy. Their script tells the story about the occupation of Carpenters Estate by the Focus E15 Mums, a group of young single mothers threatened with eviction, who ran the tower block as a social center for two weeks. This play is about preserving the history of this struggle and sharing experience with other tenants in London that fight housing regeneration. The second essay, 'Against disappearance: A partial cartography of Vila Autódromo's resistance against eviction,' seeks to underline care as a political practice in the art of housing struggle, by elaborating care and its capacity to disobey power. Cristina Ribas and Lucas Ico's reflection on their project *Vocabulary in the Movement* was built with the residents of Vila Autódromo, who have been resisting eviction in Rio for many years. Their writing of the insurgent history reveals care as the practice that gives tissue to collective structures in the coexistence of self-care, care for the family, care for the neighborhood, and care for surroundings and nature.

The first text in the second section, *Fighting Artwashing and Building Tenant Power* is by the Southwark Notes group from south-east London, which has been organizing and writing against gentrification in their area for more than 20 years. In their essay, they give an insight into the struggles against social cleansing, including artwashing, from the messy perspective of the struggle itself. In their contribution, they ask difficult questions and look at the complicated decisions made across many local groups and campaigns in the fight against artwashing and for the right to a home. This is followed by a contribution from the Los Angeles School of Echoes, 'Flipping the script on artwashing: Fighting gentrification with tenants' power,' a multi-racial and cross-generational collective of organizers, teachers, and artists dedicated to tenants' struggles in their hometown. They tease out three scripts used in art-oriented development and consider action to flip these scripts through tenant power, based on their own experience on the ground. This is followed with the text 'Hegemony and Socially-Engaged Art: The case of project Row Houses, Houston, Texas, USA' by Alyssa Erspamer, which starts from the marginal to modernity concept(s) of cultural hegemony to discuss the Raw Houses social art project created by US artist Rick Lowe. Erspamer tackles artwashing mechanisms for the current hegemony and counter-hegemonic potential of art.

The section *Politics of the Past and Politics of the Future* continues with the article 'Complicities, solidarities and everything in between. Art and political engagement in the housing movement in Bucharest and Cluj.' Ioana Florea and Veda Popovici write from the complicated position of activists, artists, and theoreticians about walking the thin line of solidarity with the dispossessed in housing struggles, mostly Roma people, and their (un)wanted connection with the real-estate market. By focusing on the post-1989 period and the emergence of a new class of restitution landlords, they reveal how community art projects prepared grounds for an enduring housing movement still active today. 'Art in the Interim: How the issue of the restitution of housing in unified Berlin led to an artistic reimagination of the city' develops an important insight into the phenomenon of interim housing use during the periods of crisis and the so-called transition in post-socialism. In this essay, Nicola Guy looks into vacant buildings under restitution used by the producers of autonomous and institutionalized art exhibitions, after the Berlin Wall fell in 1989.

Guy seeks to understand the impact of restitution on the art scene and ways in which art has been utilized both in the real-estate industry and as a tool for the city of Berlin to reimagine itself.

The section *Narrowing the Space and Mourning the Losses* starts with a contribution by Josephine Berry and Anthony Iles: 'The Exploitation of Isolation: Urban Development and the Artist's Studio.' They examine the five stages of the artistic studio's genealogy: from the isolated studio, the factory studio, open or community studio and the networked studio, to the pop-up studio. They show how the forms have developed both in relation to and at a distance from the housing market and real-estate industry. With this contribution, the authors show how housing and studios have been merging into live-work arrangements, making and remaking artists as potentially rebellious (and revolutionary) subjects that have tried to negotiate, make compromises, and eventually become the often-unwilling accomplices of social cleansing. The section ends with a contribution by Sylvi Kretzschmar exploring the potentials for art in dealing with loss in housing struggles. Kretzschmar reflects on the Morning Machine performance, which features twelve women with megaphones amplifying a set of interview-statements of former residents at the demolished Esso Houses estate in Hamburg. Kretzschmar writes about the necessity of amplifying grief in dealing with the failures of the struggle as a means of empowering people in the movement.

The book ends with a section on *Patterns of Commoning and Working with Dirt*. Here, Andreea S. Micu reflects on the role of art in Metropoliz, a squat on the periphery of Rome that provides a home for over three hundred people, mostly immigrants from Morocco, Tunisia, Eritrea, Sudan, Poland, Peru, the Dominican Republic, Ukraine, Romania, and for working-class people from Italy. She examines the role of art in the commodification of urban commons by looking at the MAAM (Museum of the Other and the Elsewhere). In looking at the Metropoliz-MAAM relation, she redefines commons as radical differences and irreducible plurality. This is followed by the text written by Filip Jovanovski, Ivana Vaseva, and Kristina Lelovac, elaborating on the histories of struggle over the site of the Railway Residential Building, a socialist housing estate in Skopje. In this script, the authors explore the problems of dwelling together and managing common spaces during the Yugoslav, anti-Yugoslav, and post-Yugoslav period. The book ends with a provocation by AHHAAHHA that raises essential issues about knowledge production in academia tackling housing violence. This contribution raises a point about the need to find ways to produce knowledge that will allow for the messiness of our lives to speak up and make new institutional arrangements against the class, race, and gender divide.

In this publication cultural theorists, activists, organizers, and artists engage with struggles and art practices to provide building blocks for understanding radical housing. In their writings, they respond to the imperative to change how we do housing and how to practice art in difficult situations when life is at stake, including one's own. This kind of messy art practice demands messy theory and a messy process of thought.[38] But beyond advocating messy knowledge production against formalized academic ways of thinking, a few writers also argue for creating the conditions for producing knowledge from struggle and from dirt. From development-led art

---

38   Elena Marchevska, 'Maternal Art Practice – An Emerging field of Artistic Enquiry into Motherhood, Care and Time' in Elena Marchevska and Waleria Walkerdine, *The Maternal in Creative Work – Intergenerational Discussions on Motherhood and Art*, London: Routledge, 2020, pp. 3–4.

to radical housing art this book advocates staying with the difficulties, namely, thinking through the existential and material contradictions of artists and tenants trapped between the struggle for decent housing and the exceptional requirements of artistic production.

References

Amin, Ash and Lancione, Michele (eds.) *Grammars of the Urban Grounds*, Durham: Duke University Press, forthcoming.

Bathla, Nitin and Garg, Sumedha, 'Radical housing and socially-engaged art: Reflections from a tenement town in Delhi's extensive urbanisation'. Radical Housing Journal 2, no.2 (2020): 35—54.

Brooks, Maya. 'Art Advocacy: The Aesthetic of Solidarity',*Circa: The NCMA Blog*, 23 Jul 2020, [Accessed: 23 Jun 2021].

Buden, Boris. 'It is "Now", Comrades. On Gal Kirn's Partizanski prelomi' in *Slavica Tergestina: European Slavic Studies Journal* Vol 17 (2016): 160–171.

Edwards, Zena. 'A Photo of a Girl – Tribute to the Ogoni 9' in African Writers Abroad and PLATFORM (eds.) *No Condition is Permanent, 19 Poets on Climate Justice and Change*, PLATFORM: London, pp. 40–41.

Kirn, Gal. 'The Yugoslav Partisan Art: Introductory Note' in *Slavica Tergestina: European Slavic Studies Journal* Vol 17 (2016): 9–17.

Lancione, Michele. 'Radical Housing: On the Politics of Dwelling as Difference'. *International Journal of Housing Policy* 20, no. 2 (April 2020): 273–289.

LATU https://latenantsunion.org/en/ [Accessed: 29 Jun 2021].

Marchevska, Elena. 'Maternal Art Practice – An Emerging field of Artistic Enquiry into Motherhood, Care and Time' in Elena Marchevska and Waleria Walkerdine, *The Maternal in Creative Work – Intergenerational Discussions on Motherhood and Art*, London: Routledge, 2020, pp. 3–4.

Milohnić, Aldo. 'Radikalni amaterizam', *Priručnik raškolovanog znanja*, Beograd and Skoplje, TkH and Kontrapunkt, 2012.

Močnik, Rastko. 'The Partisan Symbolic Politics' in *Slavica Tergestina: European Slavic Studies Journal* Vol 17 (2016): 70–85.

Murawski, Michał. 'Introduction: Crystallising the Social Condenser', *The Journal of Architecture 22*, no. 3, 2017, pp. 372–86. and Michał Murawski and Jane Rendell, 'The Social Condenser: A Century of Revolution through Architecture, 1917–2017', *The Journal of Architecture 22*, no. 3, 2017.

Muller, Martin. 'Footnote Urbanism: The Missing East in (not so) Global Urbanism in Michele Lancione and Colin McFerlane (eds.), *Global Urbanism: Knowledge, power and the City*, London: Routledge, 2021, pp. 88–95.

Patricia DeRocher, *Transnational Testimonios: The Politics of Collective Knowledge Production*, University of Washington Press: Seattle.

Sholette, Gregory. *Delirium and Resistance: Activist Art and the Crisis of Capitalism*, London: Pluto Press, 2017.

Sekulić, Aleksandra. 'Pojam radikalnog amaterizma i mogućnost kontinuiteta posle raspada Jugoslavije', *Beton* No 179, 2017.

Stakemeier, Kerstin and Vishmidt, Marina. *Reproducing Autonomy: Work, Money, Crisis and Contemporary Art*, London: Mute Publishing, 2016.

Stalder, Felix and Sollfrank, Cornelia. 'Introduction' in Cornelia Sollfrank, Felix Stalder, and Shusha

Niederberger (eds.), *Aesthetic of the Commons*, Zurich: Diaphanes, 2021: 11–38.

Stojić Mitrović, Marta and Vilenica, Ana. 'Enforcing and Disrupting Circular Movement in an EU Borderscape: Housingscaping in Serbia'. *Citizenship Studies* 23, no. 6 (18 August 2019): 540–558.

Šuvaković, Miško. *Istorija umetnosti u Srbiji – XX vek: Radikalne umetničke prakse*, Beograd: Orion Art, 2010.

Ultra-red. 'A Response to "Op–Ed: An Ultra-red Line"', X-TRA, 12 October 2017, https://www.x-traonline.org/online/ultra-red-a-response-to-op-ed-an-ultra-red-line [Accessed: 25 May 2021].

Vickery, Jonatan. 'The Emergence of Culture-led Regeneration: A Policy Concept and its Discontent' (Research papers) University of Warwick: Center for Cultural Policy Studies, 2007.

Vilenica, Ana, Mentus, Vladimir, and Ristić, Irena. 'Struggles for care infrastructures in Serbia: Dispossessed care, coronavirus and housing', Special issue: Caring in Times of a Global Pandemic, (eds.) Emma Dowling, Birgit Sauer, Ayse Dursun, Verena Kettner und Syntia Hasenoehrl, *Journal for Historical Social Research*, forthcoming in 2021.

Vilenica, Ana. 'The Doomed Pursuit of Dignity: Artists as Property Guardians in and Against Artwashing', *AM Journal of Art and Media Studies,* forthcoming in October 2021.

Vilenica, Ana. 'The Art of New Class Geography of the City: Culture-Guided Urban Regeneration Serving the Modernization of the Periphery' in Gordana Nikolić and Šefik Tatljić (ed.), *Grey Zones of Creativity and Capital*, Amsterdam: INC, 2014, pp. 51–64.

Vilenica, Ana. 'Becoming an Accomplice in Housing Struggles on Vulturilor Street', *Dialogues in Human Geography* 9, no. 2 (July 2019): 210–213.

Vuksanović Macura, Zlata. *Život na ivici: Stanovanje sirotinje u Beogradu 1919–1941*, Beograd: Orion Art, 2012.

Wright, Stephen. *Toward a Lexicon of Usership,* Eindhoven: Van Abbemuseum, 2013.

LAND OF THE THREE TOWERS

YOU SHOULD SEE THE OTHER GUY

This play tells the story of the Focus E15 Occupation of four empty homes on Carpenters Estate, Stratford, in 2014. Every word in this script, including the songs, is direct verbatim spoken or written by Carpenters Estate residents and housing campaigners. The words are taken from documentary footage by Fran Robertson, interviews by Emer Mary Morris and media news stories from the time, as well as letters and diary entries. We have stayed true to the way in which the words were said and have kept in grammatical errors and accents. CJ's verbatim was recorded later, during the devising process. Any company wanting to stage this show may choose to replace CJ's scenes with another contemporary example of housing injustice. If you do put on this play, please remember the intention is to celebrate these words, through song and collective playfulness, while staying true to them. Land of the Three Towers is a truly collaborative effort and wouldn't be possible without the tireless work of campaigners, friends, supporters who fiercely believe in the potency of telling these stories. This play is dedicated to all the fighters for Carpenters Estate, Stratford. Especially to Focus E15 Campaign and the residents who we've lost: Mary Finch and Eddie Benn.

The fight for the Land of the Three Towers continues.

KEY CHARACTERS:

JASMIN STONE: A core organiser from Focus E15 Campaign, one of the 29 mothers who received an eviction notice to leave the the Focus E15 Hostel when Newham Council pulled the funding for their emergency accomodation mother & baby unit in 2013. Jasmin has a 3 year old daughter, Saffia, who was present throughout the occupation.

SAM MIDDLETON: A core organiser from Focus E15, one of the 29 mothers from the Focus E15 Hostel.

MUTLEY: A long term resident of Carpenters Estate, he lived on the 17th floor of Dennison Point for many years and was evicted shortly before the occupation.

MARY FINCH: A Carpenters resident for over 40 years and a key fighter to save the estate. She was a core member of Carpenters Against Regeneration Plans (CARP) and can be seen speaking about the regeneration on YouTube.

FRAN ROBINSON: A film-maker who documented the footage from the occupation. She then donated the footage to Focus E15 Campaign and it was used to make this play.

ENTHUSIASTIC REPORTER: An online reporter (our very own ensemble member, Sapphire Mcintosh) who visited the occupation.

KEVIN AND FRANCIS: A young couple who've experienced homelessness and who live in temporary accommodation. They regularly visited the occupation.

ROBIN WALES: The Mayor of Newham from 2002—2018. He was renowned in ruling the council with 'an iron fist' and has a mayorship full of controversies and scandals. Due to the work of Focus E15 Campaign he was finally deselected in 2018.

EDDIE BENN: A long standing resident of Carpenters and chair of Carpenters Tenants Management Organisation (TMO), lived on 13th floor of Dennison Point.

SAFFRON: A 14-year-old local who lived in temporary accommodation and became involved with the campaign through the occupation.

AVIAH, EMER, AYESHA, RUTH: Focus E15 Campaigners involved in the occupation.

ANDREW: A Focus E15 campaigner and member of the Revolutionary Communist Group (RCG) who was involved in the occupation.

HELEN: A campaign supporter and member of RCG, with new born baby at time of occupation.

JACK, MARK, JACOB: Activists involved in the occupation.

THAMES WATER MAN: Man working for Thames Water!

LOCAL COUNCILLOR: A Labour councillor who visited the occupation.

ENSEMBLE

SOPHIE: NARRATOR [playing herself]/ KEVIN

CJ: NARRATOR [Playing herself]. CJ has her 1 year old baby, Baby G with her on stage.

SAPH: JASMIN STONE

GRACE: SAM / MUTLEY / ROBIN WALES

MARIA: EDDIE BENN / SAFFRON / JACK

LOTTE: MARY FINCH / ENTHUSIASTIC REPORTER / THAMES WATER MAN

RUBY: EMER / FRAN/ LOCAL COUNCILOR

Some notes...

-The rest of the smaller characters are shared by the ENSEMBLE and can be moved around at the director's discretion.

-The play is very musical. Each character has their own theme song with a melody taken from the intonation and rhythm from the verbatim.

-/means the words overlap.

-The ENSEMBLE are on stage the whole time. When they are not actively part of the scene they are playing music, making tea, cleaning, organising the occupation or watching the action.

ACT 1

SETTING:

The Audience are warmly welcomed into a vibrant community centre in the middle of a London Housing Estate. SAPH, GRACE and RUBY greet them, taking them to their seats, showing them around and offering them tea and nibbles.

SOPHIE and CJ, the NARRATORS, sing a medley of songs from the play with LOTTE harmonising and MARIA playing a selection of pots, pans and other household items as percussion.

The walls are busy and colourful with multiple hand-made banners depicting slogans such as: 'Social Housing, Not Social Cleansing' and 'Repopulate Carpenters'. Bunting, festoon lighting and three large architectural drawings of the towers of Carpenters Estate (Dennison Point, James Riley and Lund Point) hang from the ceiling.

The playing space is in the centre of the room, with the audience on three sides. MARY'S arm chair sits amongst the crowd in the front row. The stage is marked by a patch of thick green grass-like carpet. The 'Free Shop' of donated clothes, toys and trinkets is packed up in boxes which are scattered across the stage. The items inside will become the costume signifiers for each character and be unpacked throughout the show, to create different settings. The stage is lit with multiple domestic lamps which are operated by the ENSEMBLE throughout the show. The aesthetic has a chaotic, childlike and DIY feel.

Once everyone is seated SOPHIE and LOTTE start playing CJs Theme Song

SCENE 1: CJ

As CJ begins speaking LOTTE and SOPHIE harmonise along to parts in bold.

CJ: I've always had a kind of weird sense of home. **Before I was pregnant** I didn't feel like my family home was my home. **Before I was pregnant** I was homeless a few times, I was homeless with my family too when I was a child, so I kind of had that thread running through, right from the off.

Before I was pregnant I booked a one way ticket to Mexico, because I was like, I'm over this country, I'm gonna leave. And when I found out **I was pregnant** I was like... wow, I can't run away from my problems essentially. A big challenge was finding somewhere to live. It's a big challenge finding somewhere to live anyway, but **when you're pregnant**, and just about to **have a baby** it's a whole new world.

LOTTE and SOPHIE start 'oohing' underneath CJ's words

When you first have a baby, you can't just go out and get a job, I mean your body is in shock. When you first have a baby you're full time, every day is taking care of another life, I mean it's touch and go at the beginning that this person is actually gonna make it in the world, especially in a city like London.

Fig. 1: Carley-Jane Hutchinson, Lotte Rice and Jennifer Daley rehearsing for Land of the Three Towers at the Carpenters and Docklands Centre, Carpenters Estate, Stratford. February 2015. Photo by Tegit Cartwright.

SCENE 2: MARY

The ENSEMBLE become a Greek Chorus and start humming MARY'S Theme Song.

SOPHIE: Mary Finch, one of the 300 residents left on Carpenters Estate.

LOTTE makes her way to Mary's chair. The ENSEMBLE gather around her, watching her calmly as she speaks, singing along to her words with the parts in bold.

MARY: I'm Mary Finch, I've lived here for 42 years, had two children, Susan and Brian. I loved working in the school, started off as a dinner lady, **it's brilliant**. I became very ill in March, And Doloris, if I'm not around for a day or two, Dolaris will knock to see if I'm fine.

When I walked into here, God, I must admit, I was over the moon. I was stuck in this place that was fallin, and I mean fallin down. So when you get here [gasps and looks overjoyed] aaaaaaah!!

It was just a house you know, it was my house. My husband Brian, his brother, he had this thing about that song, 'Nelly the Elephant' And after I came home from hospital, after my op. He threw me round the room, from one end of the room to the other, doing 'Nelly the Elephant' in a jive [*laughs*].

[*Changes tone to a seriousness and severity*] People must think now: 'Oh look at them. Living in places like that!' But they've not been inside them. Don't go by the appearance of what Mr Mayor has done! Putting those boards up.

[*Leans forward in her chair and increases in intensity*] This is what I have to say to him:

If you honestly think.

That I am going to give up

my home,

to you,

or to anybody else, at my time of life,

forget it.

I will fight you,

you will have to drag me out.

It is mine.

It belongs to me.

Lights cut out

SCENE 3: OPENING

CJ: Last night, Carpenters Estate, 11pm

RUBY becomes EMER. She stands center stage and lights a torch under her chin, providing the only light on stage. She speaks frantically and directly to the audience, in a half whisper. Underneath her words SOPHIE and CJ create 'shhhhhing' noises into the mic and SOPHIE plays a quickened, tense version of the Occupation Song on the guitar.

EMER: We're somewhere between James Riley Tower and the car park. It's just me and him, trying to look casual, but he was walking a few steps ahead of me which I thought was pretty stupid cos it didn't look like we were walking together, but whatever. We walked past this alley and a dog barked really loudly, made us both jump and we 'ssssshh-ed' it, wishing it back to quiet. It did, thankfully. Then we were at the back door. I kept look out - pretending to play on my phone, sure I was shaking a bit, blood, like, pumping in my ears. A couple walked passed and I cough so he knows it's okay. It feels like hours. And then we're in. We're in.

SCENE 4: INTRO TO OCCUPATION

Lights up. As SOPHIE speaks the ENSEMBLE congregate centre stage, smiling with glee and looking above the audience heads, watching the boards come down on the four empty flats, the doors being unlocked and the Occupation opening up.

SOPHIE: 'For real politics, don't look to parliament but an empty London housing estate' —

Adita Chakrabortys words in the Guardian spoke to me.

So I went with my daughter.

Amid boarded up flats and abandoned gardens the party continued, punctuated with cries of 'these homes need people, these people need homes'.

Then, at about four o'clock the hosts hoiked themselves up into one of the flats, prised off the boards, and invited in guests.

The mums are still inside, having converted a decent, needlessly empty home into a community centre.

SOPHIE begins the Intro to Occaption song singing 'ba da ba ba...' The ENSEMBLE speak the words of the Occupation while they unpack boxes to make the 'Free Shop', hang banners, put pictures on the wall. They build the occupation as they build the song, rhythmically layering their words and creating a soundscape from the verbatim.

FRAN: [*holding camera*] Yeah, well, what I kind of felt was, if I came down and just started doing a bit of filming, and you can see it, you, kind of, as it goes on, can decide whether you trust me or whether you don't. And then, either at a later point, you know, maybe I'm not going to make a documentary or whatever, and well, the footage will always be yours to use anyway, however you wanna use it.

LOTTE: Hello so, here we are.

GRACE: Yeah they are they're beautiful.

MARIA: Come upstairs I'll make you a cup of tea.

RUBY: Yeah, yeah it's 80-86 Doran Walk Postcode E15 2JJ.

LOTTE: Yeah, I've got another little one.

GRACE: So babe, you know that 'know your rights' workshop?

RUBY: Yeah, occupying it as a political statement, um

SAPH: Why'd they board em up? Nothing wrong with em.

MARIA: Wouldn't it be nice if people were more like my cat, wouldn't it?

GRACE: Ah greaaaat. We didn't have any plaaaaates. Amaaaaazing.

LOTTE: Sorry, should I start that bit again?

SCENE 5: SAM AND JASMIN

Music Stops. The ENSEMBLE carry on decorating the occupation in the background. GRACE becomes SAM and SAPH becomes JASMIN and they stand together centre stage.

SOPHIE: Can you tell me if you've faced any of these issues in the past 5 years? Overcrowding?

SAM: Yes.

SOPHIE: Poor conditions?

SAM: Yes.

SOPHIE: Problems with landlords?

SAM: Yes

SOPHIE: Homelessness?

SAM: Yes.

SOPHIE: Benefit changes?

SAM: [*laughs*] Yes.

SOPHIE: Threat of eviction?

SAM: Yes.

SOPHIE: Rent arrears you already told me, problems with neighbours?

SAM: Not mine personally.

SOPHIE: Bedroom Tax you already told me about.

SAM: Yeah.

SOPHIE: Insecure accommodation.

SAM: Yeah, its private rented, short hold tenancy, and it's expensive: 249 pounds a week.

SOPHIE: Where were you living before that?

SAM: Focus.

SOPHIE: Which is... in Newham?

SAM: Yeah, Stratford.

SOPHIE: And that's a hostel?

SAM: Yeah. [*Beat*] I was one of the mums who was actually living in Focus at the time of the eviction notices and stuff.../so em

JASMIN: /So em, you're only supposed to stay in Focus for 6 to 8 months, cos they say that, once your child starts crawling, it's not suitable anymore. Me and Sam used to have to go to Westfields so that my daughter, Saffy had room to run around! And I was actually there for two years (beat) and it was so stressful, so claustrophobic, it's got this really tiny space with one window, you can put your arm out and touch both walls.

SAM: Loads of people were like: 'you open your legs young'. Like, but they don't actually get the actual gist of the actual situation. We were in a hostel for a reason, you know what I mean!? Anyway, me and Jasmin just started talkin' and that's when we realised that it wasn't, like, just one of us that got it, it was all of us.

JASMIN: On the 20th of August, 2013 all 29 of the Focus mums were issued eviction notices saying we /had to leave on the 20th of October.

SAM: /had to leave on the 20th October and it was ridiculous cos I was pregnant and my due date was the 19th of October and [raises eyebrows and crosses arms] I was basically shitting it.

JASMIN: We were told to look for properties ourselves, but no one would accept us coz we was all on benefits. Some people were like, 'oh yeah, we accept DSS' then they'd say 'have you got children?' and we'd say 'yes' and they'd say 'oh no, we're not taking you'. It was so hard. We went to see the council but they just registered us officially homeless and offered us accommodation in /Manchester, Hastings and Birmingham

SAM: /Manchester, Hastings and Birmingham, because 'there's no housing in Newham' and we was like: well NO, we know there is. I'm like [pulls WTF face], do you know what I mean?? [laughs]

JASMIN: Me and Sam called a meetin' for the mums, and 18 mums came, it was brilliant. We started off a petition, it was just, like, a hand-written petition and we got all our families to sign it and stuff. And, then when we was walking up Stratford High Street to get it printed we bumped into the Revolutionary Communist Group who were handing out leaflets about the bedroom tax. We asked them how we could get more people to sign the petition, and they helped and we've been doing the Focus E15 stall with them every Saturday since. There are /loads

SAM: /loads of properties. I mean go down every single street, like, you'll at least see one boarded up property.

JASMIN: We're occupying these empty flats to raise awareness, and put pressure on the council to rehouse people, hopefully, they'll listen to us and put people in the homes /because

SAM: /because we know that if there are places like Carpenters Estate, there are loads of estates out there like this. And it's... it's... horrible. Cos you know... there's homeless people out there, who desperately need a home... and they're sitting right here.

SOPHIE: And how old are you?

SAM: I'll be 21 next month... everybody's invited.

SCENE 6: ENTHUSIASTIC REPORTER

The music quickens into a fast version of the Occupation Song. The ENSEMBLE become one organism behind LOTTE who becomes ENTHUSIASTIC REPORTER. Collectively, they rush around the stage, using flocking to stick together as one unit and getting increasingly more frantic, delighted and in awe as the scene progresses.

ENTHUSIASTIC REPORTER: [*holding a hairbrush mic, speaking with a thick Northern accent*] So! I'm in the property now, and, as you can see… it's decorated! It's got information on, talking about how maybe this is part of social cleansing rather than social housing. So, we're gonna go upstairs and just see how actually really liveable it actually is.

So, I'm in one of the bedrooms now, and as you can see [*waving arms around*] it's a decent size [*stretch arms in the air*] can stretch, [*jumps*] can jump, er, got a bedside table, we've got a wardrobe, fancy playing on the guitar? [*mimes the guitar*] got a guitar! Toys! Room for books, in the corner there we've got a TV! I mean… what more… what more could a young child want!?

Here I am in the kitchen! I mean we've got space for an oven, we've got sideboards, we've got cupboards, there's food in there, look at that!! Food for a family!! What more could you ask for!? Around here we have got a cupboard! I know, I think my mum used to leave the ironing board in there, there's a boiler! It's fantastic!! Radiator!! A ceiling!! A REAL WALL!

The ENSEMBLE fall on the floor, exhausted.

I mean is this just me or is this not, is this liveable!!?

SCENE 7: LETTER TO BORIS JOHNSON

Music slows and becomes something that resembles a bitterly cold day. The lights dim. The ENSEMBLE pick up hats, scarves and placards from the Free Shop and huddle, freezing and angry, around RUBY, who becomes CARPENTERS RESIDENT.

CJ: Open Letter to Boris Johnson from the Residents of CARPenters Estate, December 19th 2012.

CARPENTERS RESIDENT: Dear Mr Mayor,

Residents of Carpenters Estate are extremely disappointed that you have broken your promise, thrice, to come and see us on the Carpenters Estate.

What was even more frustrating, was queuing for hours outside Stratford Old Town Hall on the 12th December in the freezing cold, and only allowed to ask one question, which was to ask what you intended to do about helping us to save our homes? You said that this was something that we would have to take up with Newham Mayor, Robin Wales.

As you have said in the past that you are opposed to the type of 'Kosovo style' social cleansing happening in London. Here is a golden opportunity for you to prove that you're not just full of hot air, and actually use your influence to save us from losing our homes.

Please don't wash your hands of us, like Robin Wales.

ENSEMBLE: We are the forgotten people of the Olympic Legacy.

SCENE 8: SAFFRON

MARIA throws off her hat and scarf and becomes SAFFRON.

SAFFRON: Stratford used to be the heart of our area/

NARRATOR: /Saffron, 14, Newham council are trying to move her mum to Birmingham.

SAFFRON: [*Listing on her fingers*] Stratford... Canning town... Beckton... What we should show people is what East London used to be, it used to be a family place, it used to be people for people, and how it is now... we're not even able to stand on our own two feet! [*Talking to someone offstage*] If you get me a poster about the open house, I'll put it up in my school, coz four of my mates have been moved to birmin'am, so I know quite a few people in that school that's been goin through it.

SCENE 9: HOUSE MEETING

The ENSEMBLE huddles together and collaboratively mime making a cup of tea in an increasingly elaborate and exaggerated way. Each line is spoken quickly and sharply.

EMER: House meeting!

JASMIN: We've got so much bread

EMER: We've got so much bread

JASMIN: We've got loads of bread

SAM: Who's on sandwich duties?

AVIAH: Contacting, networking, more people occupying.

EMER: Free shop, show flat

AVIAH: Security!

ENSEMBLE: Get a big finger up on em!!

JASMIN: Storyboard: carpenters residents putting up their stories

SAFFRON: Storyboard: Stratford — from family homes to posh upper class flats.

EMER: BBQ, Free shop.

SAFFRON: How long are yous lot actually planning to stay here?

ENSEMBLE pause

  EMER: Well… till it finishes

ENSEMBLE continue

AVIAH: Media

JASMIN: Who we've had

AVIAH: Who haven't we had

EMER: Who do we want who we haven't had?

AVIAH: The mirror

SAM: Big Issue

AVIAH: Press statement?

EMER: Lots of people wanting to make documentaries

AVIAH: Gotta be careful though coz they'll tell you it's called community space and it's actually called benefit street or whatever.

ANDREW: 'Gypsies on benefits and proud' as if they're not oppressed enough.

SAM: We got a lotta articles

EMER: [*Remembering humorously*] Evenin' Standard

ANDREW: 'Agitators and hangers on'

EMER: No media in show-flat, need this space to make a cuppa and not worry about dirty tea bags on the side they can zoom in on. Jobs all round? Meeting adjourned!

The ENSEMBLE tidy up the mess of tea bags, sugar and milk they've made.

Fig. 2: Lotte Rice, Ruby Wild, Sophie Williams, Maria Hunter, Sapphire Mcintosh, Carley-Jane Hutchinson and Grace Surrey performing Land of the Three Towers at The Rotunda Hall, Cressingham Gardens Estate, Tulse Hill. March 2016. Photo by YSSTOG.

SCENE 10: CJ'S PERFECT HOUSE

SOPHIE: So, what would your perfect living situation be?

CJ: Ideally I want a:

ENSEMBLE: [Sing along like a punk song while tidying]

Two bed!

Nice, like!

Nice build!

Low rise! x3

CJ: ...in East Dulwich

ENSEMBLE continue singing as CJ shouts over.

CJ: Near a really good school, erm...

Where there's a communal garden, and...

Where we can have a dog!

Maybe people around to help look after G so I can

Get a job.

And we can park a car...

ENSEMBLE stop singing suddenly.

CJ: That's just too much to ask for isn't it?

SOPHIE: And how does that compare to what you have now?

CJ: Not that at all. Can't even compare that to that.

SCENE 11: EDDIE BENN

MARIA puts on coat and glasses and becomes EDDIE BENN.

SOPHIE: Eddie Benn, lived on Carpenters for 40...5 years. Chairman of the Tenants Management Organisation.

SOPHIE presses play on an old stereo, which plays a recording of EDDIE BENN. MARIA lip syncs along to his voice.

EDDIE: It's impossible to second guess London Borough of Newham. 40...5 years?

FRAN: [heard on the recording] On the estate?

EDDIE: Yes. We had come from a truly run down tenement...Opened the door and my wife burst into tears, because it was abso- you know, it seemed like 'eaven at the time. well you're halfway to 'eaven when you're in on the 11th floor [*laughs*] ... erm... their original promise was that, um, when the place gets redeveloped, anybody that anyone who was moved off would have the right to come back.

Yeah... sure, sure.... Ain't gonna happen.

People who should have been doing things haven't, and the girls in there [*points to occupation*] have opened up their eyes.

SOPHIE stops the recording.

SCENE 12: WATER TURNED OFF

SOPHIE: 26th September 2014, the council arrive early in the morning and turn off the water to the occupied flats.

SOPHIE plays a tense version of the occupation theme in the background. Throughout the scene the Time Stamp is called by a member of the ENSEMBLE and spoken directly to the audience.

ENSEMBLE: 9.30am

EMER: [Carrying in water bottles] A police van, the council and a guy with some tools coming round, they switched off the water, they tried to switch off the leccy too/

SAM: [*Looking out the window*] /Thames water are here! I need someone to come down on the door!

MARIA: What's that?

SAM: I need someone on the door!

JASMIN: Watch out it might be a trap!

ENSEMBLE: 9:35am

JASMIN: [*Walks in with THEMES WATER MAN*] Thanks for coming.

THAMES WATER MAN: Don't start thanking me, if it's on your property I can't touch it, we can only touch the footpaths.

SAM: It's on the footpath.

JASMIN: Is it legal to turn off the water?

THAMES WATER MAN: Errm

ENSEMBLE: 9.45am

THAMES WATER MAN: Can't touch it.

SAM: How come you can't touch it, it's on the footpath.

THAMES WATER MAN: Nah nah nah, right, [*moves back a foot*] see where I am, this is what we call public footpath. [*He steps to the side*] this belongs to the council. And that's theirs. You see this one, this W that means it's Thames Water. If they'd turned that off I could have switched back on. I can't touch it, I'm sorry. But, I tell you, whatever they've turned it off with, they've pulled something out. There's nothing in there.

SAM: They've deliberately broke that!?

THAMES WATER MAN: Let's put it another way... it's broken. I don't know who's broke it... but it's broken.

SCENE 13: COURT PAPERS

CJ: Two hours later. A court porter tries to serve court papers to Jasmin outside the occupied flats.

The music quickens. Throughout the scene the Time Stamp is called by a member of the ENSEMBLE and spoken directly to the audience.

ENSEMBLE: 12pm

SAM, BEN, JACK and JASMIN run in and slam the door behind them. Everyone is breathing heavily.

SAM: [*Panicked*] Is it locked? /Is it locked?!

BEN: Yeah yeah it's locked.

JASMIN creeps towards the window.

SAM: Get away from the window!

JACK: Yeah, get away from the window!

SAM: Me and you cannot be seen at that window

JASMIN: Ok...

SAM: [*Laughing*] Phew.

JASMIN: [*Laughing*] I really hurt my ankle.

SAM: Yeah I know, I was running over the fucking door forgetting the fucking step was there...

ENSEMBLE: 12:05pm

BEN: Is he gone?

JASMIN: Ah, think he's still there...

WOMAN THREE: If it's a civil order then it might be a good idea to take it.

HELEN: [*Bouncing a baby on her shoulder*] I'm just saying that we don't know what the court order is for. If it's for a criminal offence for breaking and entering, then, then we don't want it. You need to check with someone who knows. Tell him to come back at 2 when we've had some legal advice.

BEN: [*Heading to the window*] Shall I...shall I say that?

JACK: What, come back at 2? No no, I don't think/

HELEN: /does anyone have any suggestions of where we can get legal advice from?

ENSEMBLE: 12:08

JASMIN: I really need a wee I'm gonna go now.

SAM: No flush remember.

JASMIN: Oh shit...

ENSEMBLE: 12:09

JACK: I just called GBC and they just said to leave them be, leave them there and ignore him [*Breathes heavily*] regardless he's going to try and get them through the door.

JASMIN: Yeah

JACK: We gotta make sure he can't, that they don't get through the door.

BEN: I feel bad for him.

ANDREW: He's an instrument of the state mate, don't shed any tears

ENSEMBLE: 12:11

JACK: He's got them through the door at the top!

SAM: Ah fuck /wait, wait front door?

BEN: Push em out!

SAM: Nobody touch em nobody touch em with your hands.

The ENSEMBLE tiptoe to the door and looks at the papers from a distance,

terrified.

ENSEMBLE: 12:15pm

JACK: [*Back to group*] Everyone! this is really important, we can't put on social media we've been served coz then we're served.

HELEN: Ok

JASMIN: Ok

HELEN: Make sure everyone knows.

ENSEMBLE: 12:20pm

WOMAN THREE: [*Enters hanging up the phone*] He said, since they got it through the door, we can't pretend it doesn't exist, we should read it. Because it's been through the door it's a civil order.

HELEN: Ah, civil

WOMAN THREE: A civil court order

HELEN: Ah, civil

WOMAN THREE: Yeah, and said don't worry it's not a criminal thing at all, it's a civil thing

JASMIN: Ah civil

HELEN: [To JACK, walking in] Did you get that?

JACK: No

HELEN: Advice from legal person

JACK: Hmm

HELEN: Its civil

JACK: Ah civil

HELEN: We can stop pretending it doesn't exist

JACK: Stop pretending, ah sweet

ENSEMBLE: 12:25

JACK: [*Getting Emer's attention*] Court is today 2pm! Interim Possession Order... which means they can be in immediately.

EMER: Fuck.

SAM: [*Enters, eating chocolate*] What, what, what??

EMER and JACK turn to SAM.

JACK: Court is TODAY! At 2 pm, TODAY, today's the 26th yeah?

EMER: Yeah

JACK: [Reading the papers] Served before 11am on the 28th.

SAM: Are they allowed to just scribble out dates?

EMER: I dunno, the actual date under there says... guys! The original dates on this says it was meant to be served before 10am on the 7th of October, 2014 and someone has come along with a red pen and scribbled it out and put it today.

SAM: Are you serious?

WOMAN THREE: This is an Interim Possession Order which means that they can barge in and chuck everybody out/

HELEN: /we've got to have a strategy then, if you are going to the court

JASMIN: Yeah

JACK: So, interim possession order today

SAM: Yeah

HELEN: Ok, people need to do a massive call out – contact everyone, everyone.

ENSEMBLE: 12.30pm

SAM and JASMIN speak directly to ANDREW'S camera phone.

SAM: So please if anyone has any available time, we need to get as many people down as possible.

SAM: Thank you.

JASMIN: Thank you.

ANDREW: K, do you want me to play it back to you?

SAM: No, just upload.

ENSEMBLE leave the stage.

SCENE 14: COURT 1

NARRATOR: An hour and a half later, outside Bow County Court, Stratford.

The scene becomes underscored with the ENSEMBLE clapping in the rhythm of the Focus E15 chants. Each actor has an individual rhythm which build up and overlap as the scene progresses.

JASMIN: [*On mic*] We are the Focus E15 mothers and we have been occupying, politically occupying a flat on Carpenters Estate, that has been left empty for years and years on end, and we're here fighting for social housing. Social housing is a basic human right, everybody deserves a place to live. Come and stand with us, stop evictions, stop people being forced out of London. [*Starts to chant in the mic*] SOCIAL HOUSING IS A RIGHT

ENSEMBLE: HERE TO STAY, HERE TO FIGHT.

JASMIN: SOCIAL HOUSING IS A RIGHT

ENSEMBLE: HERE TO STAY, HERE TO FIGHT

JASMIN: [*On mic*] We've been told to come to the county court at Bow, and we are saying Newham council are the criminals, they're pushing people onto the streets, they are pushing people into private rented accommodation. We need to stand together. Look at the support we've got here, we've got people stopping, we've got people holding banners, we've had less than two hours to organise, to come here and stand outside this court today to say: SOCIAL HOUSING IS A RIGHT

ENSEMBLE: HERE TO STAY, HERE TO FIGHT

JASMIN: [*Puts mic down*] My hands are shaking.

JASMIN: [*Starts speaking more intimately, directly to the audience*] We are going to go in and stand for our rights, and fight for decent social housing. We have no idea what to expect, we are quite nervous, but we are doing this for all the people that are suffering with a housing problem, and we feel so passionate about it, and we are gonna stand in there, and we are gonna [has to take a second- as tears well] we are gonna stand in there and we are gonna fight for decent social housing for everyone... Sorry... [*takes a break, and continues*]… it's really emotional coz there are so many people suffering everyday, and this is the kinda thing that needs to happen, people organising and getting together, people fighting for a basic human right [*laughs*] I'm gonna go over there now…

JACK: Yeah, yeah, um Lindsey Jonson of Doughtery chambers, a pretty famous housing barrister, has come down to represent us pro bono. He took a taxi over soon as he heard, and in less than two hours we've managed to assemble quite a legal team. So yeah at this moment in time we feel we are is safer hands than we did about two hours ago [*laughs*].

AYESHA: [*On mic*] We are having a demonstration outside the court, because two of our comrades, two of our brave comrades are inside the court, why are they inside this court? For standing up for themselves. They are standing up for their communities, in this Olympic borough. 9 billion pounds, was spent during the Olympics on those two weeks, 9 billion pounds, and where is the Olympic legacy? Go and look on the Carpenters Estate, right next to the Olympic stadium, and what you will see there.

The ENSEMBLE tense, waiting for SAM and JASMIN to come out of court. When they do, holding up arms, fingers stretched out for victory, there is a huge cheer.

AYESHA: They've been in court the brave people from Focus E15 Campaign. Make. Some. Noise!

ENSEMBLE: [*Shouts and cheers*]

JASMIN and SAM are interviewed to camera.

SAM: We knew nothing, and, like we didn't have time to give a statement in, and basically the judge was like, you know what, because it was an unfair hearin' date, court adjourned until a further date. That means we can enjoy the weekend! [*they laugh*]

JASMIN: So that means we've got up until Thursday that we can continue raising awareness and let more and more people know about the situation, and what's going on/

JASMIN: /Beamin'… We feel amazing, it's a part victory/ and we're gonna go all the way

SAM: /Yet again part victory, but we are keep fighting, that's what we are gonna keep doing/

JASMIN: /we're not gonna stop/

SAM: /We're not gonna give up

JASMIN: We're gonna take a march back to our place with all these lovely people that have been supporting the campaign.

SAM: Yeah, yeah we've got an open mic night tonight, so you know everyone is welcome!

Fig. 3: Baby G, Carley-Jane Hutchinson and Sophie Williams as the Narrators. Photo by YSSTOG.

OPEN MIC INTERLUDE

MARIA introduces the Open Mic night. In which people from the audience are invited onto the stage to sing songs, read poetry or say whatever they'd like about their estate.

The Open Mic ends with the ENSEMBLE singing 'Like, You Know' a celebratory song compiled of different pieces of verbatim from Land of the Three Towers — some from Act 1, which the audience may recognise, and some from Act 2, which is still to come.

ENSEMBLE: Like, you know. Like, like you know. Like, like, you know. Like like, you know x2

MARIA: To build

GRACE: We'd be happy with a studio flat for like, not even a quarter of that. Like, you know. What you've done, what you do. Like, unable to.

ENSEMBLE: Like, you know. Like, like you know. Like, like, you know. Like like, you know x2

MARIA: Someone's gonna bring a Lego campervan.

ENSEMBLE: Someone's gonna bring a Lego campervan.

MARIA: Someone's gonna bring a Lego campervan [*the ENSEMBLE clap three times*] to build.

ENSEMBLE: Like, you know. Like, like you know. Like, like, you know. Like like, you know x2

CJ and SAPH: [*Singing over the ENSEMBLE*] We've been running a social centre. On Carpenters Estate. with homes that have been boarded up. For years and years on end. To the c, to the c, to the c...

ENSEMBLE: Woahhhhhhh!

CJ and SAPH: County Court!!

ENSEMBLE: Like, you know. Like, like you know. Like, like, you know. Like like, you know x2

RUBY: Opened the door and my wife burst into tears, because it was abso- you know, it seemed like 'eaven at the time. well you're halfway to 'eaven when you are in on the 11th floor [*laughs*].

ENSEMBLE: Halfway to heaven on the 11th floor

ENSEMBLE: Like, you know. Like, like you know. Like, like, you know. Like like, you know x2

LOTTE: It's livened the place up again. They're lovely and I'm quite close to you. No problem, I can see you out my window nooooo problem. If I'd been in a better mood I would have given you a few songs on the old karaoke. No problem.

CJ: If you can't afford to live in Newham, then you can't afford to live in Newham.

ENSEMBLE: If you can't afford to live in Newham, then you can't afford to live in Newham.

CJ: If you can't afford to live in Newham.

ENSEMBLE: Then you can't afford to live in Newham.

MARIA: To build.

ACT 2

SCENE 15: SOPHIE and CJ

SOPHIE starts playing CJs theme on the guitar as CJ starts speaking. LOTTE and SOPHIE harmonise over the words in bold.

CJ: Ok, when I was pregnant. I tried to move home but things went wrong. So they put me in emergency accommodation. **6 months max: 3 months pregnant, 3 months newborn.** No visitors. So they put us above a noisy music venue on a main road. Up loads of flights of stairs. Which I can't manage on my own. **Really difficult with a pushchair. Really difficult to get up the stairs.** On my floor there are seven families, next door is a family of six.

SOPHIE: In one room?

CJ: In one room

SOPHIE: Jeez, didn't know that was legal.

CJ: It's not legal, It's illegal.

SOPHIE and LOTTE start 'oohing' underneath.

CJ: They know it's illegal but they say they don't have any temporary accommodation. Legally, I can only be there for 6 weeks with a child. They said, to give you an idea, people have been waiting for 5 years. I said, well am I gonna be in a bed and breakfast for 5 years then? And they said, well there's a housing crisis, blah blah blah...

SOPHIE and LOTTE replace their 'oohing' with 'blahing'.

SOPHIE: So the council aren't planning on moving you and baby G from the bed and breakfast?

CJ: No, but they will. Cos I'm not living there. So something's gonna happen.

SOPHIE and LOTTE: [Singing] Something's gonna happen.

SCENE 16: MUTLEY

GRACE puts on MUTLEY'S hat and stares out the window, admiring his view

CJ: Mutley, lived in Dennison Point, Carpenters Estate, for 30 years. He was evicted just before the opening of the Occupation.

MUTLEY: [*Talking, as if to himself, enjoying the nostalgia*] 104, 17th floor, see the world from Mutleys... hahaa... [*shakes head*] mad. Had some great great times in there... for years and years I didn't have any feeling for it, but obviously, not too far down into your psyche... nah coz there used to be a saying around this place? Canning town, the rough lie down, forest gate for the ladies [*laughs*]

[*Responding to Interviewer offstage*] Carpenters? It was known as the land of the free towers, that's it, land of the free towers... coz Dennison you see, no matter whether you are on the underground, coming out of the tunnel, or going back into it, it's either the first thing or the last thing that you see, and on great eastern, that line up to Liverpool street, and out to Southend and beyond the other way, again it's the first and last thing you see, of Stratford. There's the Den.

SCENE: 17 KEVIN AND FRANCES

SOPHIE: Kevin and Francis, A young couple who visited the occupation.

SOPHIE becomes KEVIN and joins MARIA as FRANCIS on stage. The ENSEMBLE sit on the floor looking up at them and start singing. The harmonies develop into a choral version of 'I'm still in love, with you Boy' by Sasha and Sian Paul. As they speak, KEVIN and FRANCIS never lose eye contact and are totally unaware of anyone but each other, they slow dance along to the harmonies created by the ENSEMBLE.

FRANCIS: Yeah, we was homeless for a couple of years – then eventually, working with a lot of organisations, not through the council, we ended up both getting into separate hostels.

KEVIN: We've been together, like, three years now, and we're not even allowed visitors in our hostels. Yeah, I can't go to hers.

FRANCIS: It's gotta be outside… so we go to Art Galleries.

KEVIN: Tate Modern, Barbican, and just doing a lot of projects. Francis is on college courses now, but/

FRANCIS: /We just wanna relax together, you know? Not have to go out and do things all the time, just sit down and chill out.

KEVIN: And the money side of it is ridiculous. Because I found out the other day, they've paying like 247 pound a week for me to live in my hostel.

FRANCIS: And mine's 315 or something –

KEVIN: Yeah, so, the maths says it all, doesn't it, you know like you could get a one bedroom flat, or like we'd be happy with a studio flat for like not even a quarter of that. This has really opened my eyes, people think you can take someone off the street and you can put them in a hostel, and it's solved. But the government's spending X amount of money to keep me and my partner away from each other.

FRANCES: All they have to do is take the metal shutters off for us and we'd go in there.

KEVIN: We'd have done it ourselves.

KEVIN and FRANCES continue to slow dance, looking into each other's eyes as the ENSEMBLE continue singing until. When the scene ends they smile at one another and part ways.

SCENE 18: SAFFRON

MARIA becomes SAFFRON and stands centre stage.

FRAN [*Interviewing SAFFRON*]: So, do you live on this estate?

SAFFRON: Nah, I'm in the Olympic buildin', I hate livin' there, my boiler has been broke for 7 months, they've ONLY just replaced it. My toilet's broken and I'm like 'well hold on a minute, this is supposed to be a "fashionable" building?' WHY doesn't this stuff work? Sorry but the council buildings I've been in... a load of my mates live in council buildin's and their houses work fine. They are made out of brick, like when you knock on this yeah [*mimes knocking on brick wall*] you don't hear nuffink, but when I knock on my wall you hear a BANG! Coz you know it's just thin – it's just a thin layer it's not even a wall. That's why I like bricks coz if my mum and dad did get up to anything I wouldn't be able to hear it. That's your worst nightmare hearin your mum and dad or OR WALKIN IN ON YOUR MUM AND DAD! I think that is the worst thing EVER I did that once and I've never gone in her room again. I was just like 'Oh my god that's the worst picture I've ever seen in my whole entire life'.

RUBY dons a suit jacket and becomes LOCAL COUNCILLOR.

LOCAL COUNCILOR: [*Coughs into the mic*]

SAFFRON sees LOCAL COUNCILOR and changes her mood completely to rage, she runs over to SAM.

SCENE 19: LOCAL COUNCILORS VISIT THE OCCUPATION

SAFFRON: [*To SAM*] Basically, we've got a councillor on the camera chattin a load of shit!

SOPHIE: Wednesday 1st October 2014, 10am, 2 local councillors visit the occupation.

LOCAL COUNCILLOR: [*Trying to settle the crowd*] We're here to represent. We're... we're a collective organisation, aren't we. There's 61 of us, we've got an executive mayor model, which is erm, a very powerful position.

SAFFRON: And who's that then?

LOCAL COUNCILLOR: It's erm... Robin Wales

The ENSEMBLE react in different ways, laughing, becoming more tense, shaking their heads.

SAFFRON: Hahahahahhahaaa!!

JASMIN: Sorry... can I just say [*everyone is talking and she calmly attempts to get their attention*] Sorry... Robin Wales said to us, personally, and we have evidence of this, Robin Wales said: 'If you can't afford to live in Newham, you can't afford to live in Newham'.

RUTH: [*Very sternly*] He also said 'Don't raise it with me, raise it with the government'.

LOCAL COUNCILLOR: Well, yeah, I'm saying that, in a different way. Erm, we can make representations, as your representatives, about what is the right and proper thing to do... These buildings over here... We'll, represent/

SAFFRON: /Rep.re.sent, Rep.re.sent

JASMIN: What are you going to do about Carpenters Estate? There are properties on this estate that have been /empty for over 8 years

SAM: 8 years ago!

LOCAL COUNCILLOR: There was a regeneration/ plan it fell through yeah

SAM: /Gentrification

RUTH: /Do you believe in council housing?

LOCAL COUNCILLOR: Yes, I do.

RUTH: Then save this estate.

LOCAL COUNCILLOR: No, no, no, no the issue is wider than just, em/ council houses

SAFFRON: /We all know the issue!

JASMIN: We/ don't want short life accommodation

SAFFRON: /We've been here for two weeks; you've come the day before we go to court! What good are yous?! /Answer that one

JASMIN: /These homes are ready to be moved into!

SAM: I'm paying 249 a week rent, private rent right, do you think that is affordable?

LOCAL COUNCILLOR: No, our economy doesn't work properly, what we need to do is change our/ economy

SAM: /What we need to do is get you people out of fucking power, and get people like us in, because its people working for the people these days/ it's not yous lots working for us, it's communities working together to get this shit done.

SCENE 20: MARY 2

LOTTE becomes MARY, puts on her dressing gown and sits on her chair. The ENSEMBLE gather centre stage and become her Greek chorus again. The lines in bold show when the ENSEMBLE harmonises over her words.

CJ: Wednesday 1st October 2014, 1pm. MARY gives a witness statement to support the campaign on their second court date.

MARY: **It's livened the place up again**. They're lovely. No problems. People say you get problems with all of this — I've not seen any so far and I'm quite close to you. I can see you out my window! I've enjoyed a couple of nights over there. If I'd have been in a better mood I would have given you a couple of songs on the old karaoke! But no, it's been quite good, quite lively. Like it used to be. That's how it is, wasn't it.

Bring this recording, get across to them at this court that what they're doing to us is horrible, absolutely horrible. My health is so deteriorating, really deteriorating, it's just worn me down bit by bit by bit. Every night I go to bed and I think, 'Get up in the morning! What way can I get at them? What way can I fight em?' Because we vote these people in and look what they're doing! Where has this country gone?

SCENE 21: LEGAL MEETING

SOPHIE: Wednesday 1st October 2014, through the solicitors Newham council offer Focus E15 an out of court deal. A legal meeting is called in the show flat in the occupation.

This scene is fast paced. Throughout the scene the Time Stamps are called by a member of the ENSEMBLE and spoken directly to the audience

ENSEMBLE: 3:11pm

MARK: We said we would like to stay until the 20th, 3 weeks from now, the council came back and said YES, we can stay if: [*marks out on fingers*] you agree to have an inspection of the property, you agree to a mutually worded joint statement with the council, AND you promise not to occupy again.

JASMIN: We can't agree to those terms

MARK: Yeah, but if we don't accept that, they are going to push for an injunction, that's serious

AVIAH: An injunction to prevent what?

MARK: To prevent any of us or anybody from trespassing on any property

SAM: You serious?

MARK: And, they could kick us out in 24 hours

SAM: You serious?!

ENSEMBLE: 3.17pm

MARK: Basically we have the choice of: staying, and promising that we won't occupy, which is not legally binding. Or going to court tomorrow and risking them pushing through a legally binding thing that would stop us from being able to occupy again.

ENSEMBLE: 3:37

RUTH: Why are the council so keen to keep us out of court?

AVIAH: Coz the media is gonna be there. BBC, Channel 4, Sky, all of them. After tomorrow they are gonna start getting bored.

MARK: Our bargaining chip is the platform, the media attention, the residents, being a strong unified voice.

ENSEMBLE: 5:23

JASMIN: Let's do it on a vote

SAM: Yeah

JASMIN: We go to court, see what happens, if we've got 24 hours, we've got 24 hours, if we've got longer we stand down the day before the next court date, do you know what I mean? Then we've left on our own terms.

SAM: Yeah like we've always done with everything else.

JACOB: Right, hands up in favour of rejecting the council's offer and going to court tomorrow.

Everyone in the ENSEMBLE stick their hands up

JASMIN: I love court!

JASMIN: [Chucks a pillow at Sam's head, she chucks it back and they both giggle]

SAM: Wait does this now mean me and Jasmin have to sit down and work out what we might have to say in court tomorrow?

JASMIN: Oh crap, I forgot that! Nah we've changed our minds coz we might have to talk! Haha!

ENSEMBLE: [Whoops and applause]

AVIAH: See you in court!

SCENE 22: COURT 2

NARRATOR: Thursday 2nd of October, 12pm at Bow County Court, for the hearing of Newham Council's new Application for an 'Interim Possession Order'.

SAM and JASMIN get out of the car and walk towards court, holding hands, immediately bombarded by the press. Each person in the ENSEMBLE has their own clapping rhythm taken from a Focus E15 chant. The claps build throughout. The press speak directly to SAM and JASMIN physically getting in their way and thrusting mics in their faces.

JOURNALIST: So, are you feeling quite confident or...?

JOURNALIST 2: So! How you feeling?

WOMAN WITH BANNER: Hello! I heard you on Five Live, you were fantastic! How you feeling?

AUSTRALIAN JOURNALIST: What time you due to go in? How you feeling?

SAM: [Whispers in Jasmin's ear smiling cheekily] Hey, shall we go round the corner and have a fag?

JASMIN and SAM, round the corner from the court, smoking cigarettes.

SAM: [Nervously, to JAMSIN] I don't want the cameras lookin at me when I'm smokin. [Pointing off stage] This one's fine but not that one or that one over there. [Excited] OH YOU FOUND MY PINK LIGHTER!! [they hug]

VOICE: SOLIDARITY SISTERS!!

SAM and JASMIN: Woooo!!

WOMAN: Breathe, breathe just breathe, everything gonna be ok.

SAM: [*To JASMIN*] That's what everyone has been telling us. Breathe, just fuckin breathe.

Voice: GIVE THEM A FLAT ON THE CARPENTERS ESTATE!

ENSEMBLE: GIVE THEM A FLAT ON THE CARPENTERS ESTATE!

WOMAN WITH BUGGY: It's fight or flight and your fighting not flighting girls, yeah? You're fighting for me and my son. I love you. With all my heart. Don't make me get emotional! Mate, don't make me get emotional. You're makin' life worthwhile, you get me? We ain't dogs, we're human beings. I'm gonna smoke out of the way I think.

JASMIN and SAM go into court.

ANDREW: [*On mic*] It is great to see so many people here today. When we have been on that stall in Stratford Broadway, outside Wilkinson [*Direct to audience*] please come along this Saturday. A shout out to Saffron and all the revolutionary young women we've come in contact with throughout this campaign. Shall we do some chanting? This is an open mic, come up, use the mic. We're here to say: SOCIAL HOUSING IS A RIGHT

ENSEMBLE: Here to stay, here to fight!

ANDREW: SOCIAL CLEANSING ISN'T RIGHT

ENSEMBLE: Here to stay, here to fight!

HELEN: Has anyone seen the Newham recorder this week? Apparently, our Mayor Sir Robin Wales, whose salary is over 80 thousand pounds a year,

ENSEMBLE: Boo!

HELEN: Has called the occupation of empty flats 'despicable'

ENSEMBLE: Shame! Shame!

ENSEMBLE: [*Chanting*] ROBIN WALES. GET OUT WE KNOW WHAT YOU'RE ALL ABOUT. EVICTIONS. HOME LOSSES. MANSIONS FOR YOUR BOSSES!

HELEN: They've just gone into court, now let's have a really big cheer so they can all hear us inside!

ENSEMBLE: WAAAAAYYYY!

SAFFRON: /[*Ridiculously screechy*] Oh my god it's Russel Brand! Russel Brand aaaaagh!!

SAFFRON: [*On mic*] WHOSE COUNCIL HOUSING?

ENSEMBLE: OUR COUNCIL HOUSING!

SAFFRON: [*On mic*] We've now been moved into a bed and breakfast which is a complete pile of shit there's no bloody plumbing, we've had our electric cut off for two weeks now. And on the tenth of October we find out whether we're gonna be moved out of London.

ENSEMBLE: GIVE THEM A FLAT ON THE CARPENTERS ESTATE!

ENSEMBLE: SOCIAL HOUSING IS A RIGHT! HERE TO STAY, HERE TO FIGHT!

Whoops and Cheers from the ENSEMBLE as SAM and JASMIN come out of court.

JASMIN: [*Reading from statement*] We are truly grateful for the support and solidarity from both the local and wider community. We have decided...to leave the occupation of 80-86 Doran Walk of our own accord by the 7th of October. Newham Council have agreed to this with no conditions.

EMER: When we were sitting there with our solicitor, and our barrister and um, we knew that we needed to go, soon,

SAM: We have celebrated a year of the Focus E15 Campaign and have tried to communicate with Robin Wales and Newham council on a number of occasions. And they have refused to listen. As a result, our political occupation was the only option to escalate our demands for social housing not social cleansing.

EMER: We made the decision we made for all the right reasons, we couldn't sustain the amazing work and we wanted to leave on a high, without any court fees or charges, no conditions or injunctions, all the logical reasons.

SAM: This occupation was not about staying indefinitely

EMER: It was coming out of the courtroom,

SAM: But our demands to Newham Council to:

EMER: Coming out to the hundred or so people that were there,

JASMIN: Repopulate the Carpenters Estate. Secure council tenancies now

EMER: Everyone doing victory signs and being like 'wayyy!'

SAM: End the decanting and evictions of the existing residents

EMER: And Jasmin and Sam read the statement...

JASMIN: No demolition of the estate

SAM: The estate should be managed by the residents for the residents. No third party or private management.

EMER: I just, I saw Saffron face, and she just welled up and I just, I felt so sad, I didn't know if it was the right decision. [*Smiles, shrugs, beat*] But then I thought, look what we've done, we didn't know anything when we started — none of us had done this before.

JASMIN: We will continue fighting for council housing and ensure decent housing for all.

EMER: For some short period of time, we made something very special. And if the council don't listen, we'll be back.

SAM and JASMIN: THIS IS THE BEGINNING OF THE END OF THE HOUSING CRISIS!!

ENSEMBLE SING: 'Our house, in the middle of our estate'

SAFFRON: They made it feel like home, they made everyone come together, and for them to leave, it's just gonna feel like an empty community again. So... it completely broke my heart when I heard that they were leavin'.

SCENE 23: LETTER FROM ROBIN WALES

The ENSEMBLE find a ROBIN WALES puppet in the Free Shop. It has been constructed by children from a sock puppet and an old suit jacket. GRACE, MARIA and LOTTE bring the puppet to life. GRACE becomes the voice of ROBIN WALES, reading from his letter and imitating his Scottish accent.

SOPHIE: A letter in the Guardian, by Mayor Robin Wales, Monday 6th October 2014.

The ENSEMBLE collectively clear their throat.

ROBIN WALES: I apologise to the Focus E15 families, but this is a London housing crisis.

Newham Council will tomorrow take back possession of four illegally occupied boarded-up flats on the Carpenters estate beside the Olympic park.

When the council took the decision — with its landlord East Thames Housing — that 30 families were ready to move from the Focus E15 Foyer in Stratford, we should have engaged with them from the start, planned how we would support their next steps and worked with them individually. Although the decision was the right one, the way both their landlord and the council initially dealt with the Foyer families was unacceptable, and for that I apologise.

The 23-acre Carpenters estate requires urgent redevelopment.

Newham's residents will lose almost £4m through the bedroom tax and benefit cap this year alone, while this government has slashed the money we receive to support them.

Here in Newham we are doing our bit.

Reversing the crisis in London's housing market will take significant investment, political courage and tough decisions. It is a challenge that this Conservative government is choosing to ignore.

Robin Wales

ENSEMBLE: [*In the tune from 'Like, You Know'*] 'If you can't afford to live in Newham, you can't afford to live in Newham'

SCENE 24: EMER AND JASMIN VISIT MARY AND BRIAN

LOTTE makes her way to MARY'S chair. SAPH as JASMIN and RUBY as EMER sit opposite her, centre stage. The rest of the ENSEMBLE sit around the edges and watch them.

CJ: Thursday 2nd October 2014, After Court, Emer and Jasmin go to visit Mary.

EMER: [*Knocks on door and waits. Door opens*] Hello

JASMIN: /Helloo

MARY: [*Overjoyed*] /Hello! [*hugs*] Caught me untidy again.

EMER: So nice to see you.

MARY: [*Giggle*] I'm in a mess.

EMER: [*Laughing*] This is our tidy

JASMIN: Lovely house

EMER: Isn't it lush/

MARY: /They're nice and big aren't they

EMER: They're amazing

MARY: So how did it... how did it go?

EMER: Well I think, well [*looks at Jasmin*] you can say

JASMIN: Em, yeah so obviously we went to court today and, we decided that we'd be leaving on our own accord on the 7th of October/

MARY: /It's the best way

JASMIN: [*Agreeing but with a hint of sadness*] Hmmm. Well, we wanted to leave on our own terms. We didn't want to be chucked out and we didn't want to burn out. But we're gonna obviously keep up with the campaign and demand what we're demanding – which is to repopulate the carpenters estate.

EMER: Yeah, because they say they're putting forty homeless families onto the estate, which is great. But what we want is long term, secure tenancies.

MARY: Oh good, you're not gonna desert us then

JASMIN: No, no, never, never

MARY: Oh I did feel a bit, you know, a bit choked up

EMER: Yeah … yeah… It's good but it's sad

MARY: It's good but it's sad [*pause*] I think he is really determined to take this place.

EMER: Well, he's not gonna get it

MARY: [*Laughs*]

JASMIN: We're not gonna let that happen. We can not. And, you know, we came here cos we saw a video you did on YouTube when you spoke about carpenters estate. And it really, really inspired us. We saw your video and we thought 'we've gotta do something'.

MARY: Thank you. You know? It, it's been truly brilliant. Cos young people today, the way they get talked about, you don't realise that there are actually some goodins out there.

EMER: [*Laughs*] Well it's the start of something.

JASMIN: Yeah, we've been saying, 'This is the beginning of the end of, em, the housing crisis' [*giggles*]

MARY: [*Laughs*] Well, wouldn't I love that.

SCENE 25: OCCUPATION OUTRO

The ENSEMBLE pack up the occupation, wiping messages from the chalkboard, taking things off the wall, packing up the Free Shop. They sing the Occupation song and punctuate it with the following pieces of verbatim.

ENSEMBLE: Got some Polyfilla?

Yeah, we've got polyfilla

Now we're just packin' up and shit like that

We even came up with a slogan which is 'Redecorate, Repopulate'

FRAN interviews JASMIN and the song continues behind them.

JAMIN: [*Smoking a cigarette*] I'm a bit saaaaad. We're going home our individual places where everybody's shut away

FRAN: Does it not feel like home at all where you are now?

JASMIN: No, I absolutely hate it with a passion, literally. You need to have that sense of community and you need to have space. And the amount they charge for it... I don't know how they could charge that much for something like that. Here Saffy, mummy needs to wash you now.

JASMIN: There's only one bad thing that's happened since I got here, and that's the potty training

FRAN: So it started, and it's not going so well now?

JASMIN: It was complete [*laughs*] I've got a lot of work to do when I get home.

FRAN: Do you feel you've achieved what you wanted to achieve?

SAM: Partially. I mean if these, like, houses do get refilled with people on a long term tenancy then yeah.

Finishes cigarette and throws on the floor.

FRAN: What's the latest that the council have said?

SAM: I think they're shitting themselves, to be honest with you, [*smiling*] that we don't do it again. It's good.

SCENE 26: BREAKING INTO BRIDGE HOUSE

NARRATOR: November 2014, Six weeks later. Candice, a young woman whom Newham council refused to house in Newham, and was told that she would be moved to Hastings, goes to her meeting with Newham council at Bridge House Housing Office, supported by Focus E15 Campaign.

The ENSEMBLE huddle behind a buggy ready to break into Bridge House Housing Office.

LOTTE: Sorry?

The following story is sung collaboratively by the ENSEMBLE.

ENSEMBLE: Yeah, yeah, yeah, yeah, yeaaaaaaaah!!!!

LOTTE and SAPH: Candice went in straight away, along with somebody else from...

LOTTE: Who I was with all day but don't know his name [*laughs*]

LOTTE and SAPH: and um,

ALL: Yeah, yeah, yeah, yeah, yeaaaaahhhhh!!!!

CJ and GRACE: So when we was trying to get in, I had the buggy there and the security put his knee in Saffia's face, so I had to pull her back, and Sam was like

ALL: Uh!!

LOTTE and SAPH: As she usually is

MARIA and SAPH· And then he was pushing Sam and saying

MARIA: 'Stop pushing me!'

ALL: Punched Sam in the face,

LOTTE: And then three other security guards came along and they were blocking the door, so I'd got on the floor and

ALL: Unclipped the bottom,

LOTTE: Sam was reaching up and

ALL: Unclipped the top

LOTTE: We like

ALL: Ripped it open,

LOTTE: There was quite a few of us

ALL: Ripping it open.

CJ and GRACE: They still didn't let us in, but then I had an opportunity, when they was like pushing Sam against the other door, so I quickly ran in, got on the floor and went

ENSEMBLE: Between his legs!

LOTTE: [*Laughing hartly*] I got through his legs!!

MARIA and SOPHIE: It was quite funny and I was really surprised that I managed to get

ENSEMBLE: Between his legs!

LOTTE and JEN: And I pulled Sam through

ENSEMBLE: And I was in!!

CJ and GRACE: Candice had an appointment and we was very persistent, and we was making sure that we was talking about Carpenters' Estate and making them

ALL: Feel. Like. Shit.

MARIA and SOPHIE: Then, after twenty minutes, they came back with a

ENSEMBLE: Two bed, Two bed, Two bed, Two beeeeeeed

Maisonette in Canning Town!!

ENSEMBLE bow and audience clap.

SCENE 27: ENCORE - WHERE DO WE GO?

ENSEMBLE: [*Singing*] Where, where, where, where do we go, go, go? x16

CJ: [*Speaking over singing*] The Focus occupation on Carpenters put an immense amount of pressure on Newham council. They opened 28 of the boarded up houses. Although only on a short tenancy – they're now homes for people who need them.

RUBY: In the months after there were more political occupations happening across

London.

MARIA: Guiness Trust estate in Brixton... Aylesbury Estate, Walworth.... Sweetsway Resists, Barnet/

SAPH: /This summer Sisters Uncut ran two occupations demanding better public services for women facing domestic violence/

CJ: /And there are many more estates facing regeneration and demolition. All over the country people are **fighting back.**

The ENSEMBLE shout the names of different estates that are at risk of regeneration over the music.

RUBY: Whitefield estate Brent Cross

central Hill Upper Norwood

West Hendon Barnet

Cressingham Gardens – Brixton

GRACE: Silchester Estate – Kensington

Fred John Towers – Waltham Forest

Sweetsway also in Barnet

Robin Hood Gardens - Poplar

And the list goes on - and on.

ENSEMBLE: [*To the audience*] Whose Council Housing!?

AUDIENCE: Our Council Housing!

CJ: In London there are more than 164,203 residents being affected by regeneration. More than 70 social housing estates and 30, 300 homes are being demolished.

Great change requires everyone to get involved.

Would you really like living in a city where everyone is the same?

And all the empty homes are second homes?

What would London look like if they got rid of everybody? Soulless. Fake. No history.

I think people will do something, but people are waiting for other people to do **something.**

The ENSEMBLE stop singing.

CJ: But the time is now, really.

Fig. 4: Artistic Directors Nina Scott and Emer Mary Morris prepare for the You Should see the Other Guy takeover of The Stratford Centre, a performance of Land of the Three Towers with talks from Focus E15 Campaign. The demo protested the numbers of homeless people sleeping rough in the shopping centre, while Carpenters Estate still lay empty around the corner. March 2016. Photo by Tegid Cartwright.

END.

Written and Directed by:
Nina Scott and Emer Mary Morris

Devised by the Ensemble:
Sophie Williams
CJ Hutchinson
Sapphire Mcintosh
Maria Hunter
Jennifer Daily
Lotte Rice
Grace Surrey
Ruby Wilde

Music by:
Ben Osborn and Nina Scott

Design by:
Nina Scott

Photo Credit:
Tegid Cartright

Supported by:
Arts Council England, Network for Social Change and The Edge Fund, Camden Peoples'
Theatre, Limehouse Town Hall, Focus E15 Campaign and many more!

AGAINST DISAPPEARANCE: VILA AUTÓDROMO'S RESISTANCE AGAINST EVICTION IN BRAZIL

CRISTINA T. RIBAS & LUCAS S. ICÓ

Since 2009, Rio de Janeiro has been in the spotlight as the next city to be exploited by mega corporations that capitalize by spreading their extractive practices across the globe. Hosting mega events such as the World Cup and the Olympics brought the combination of auto-colonization and foreign capital to the city.[1] In this way, the world-famous landscape of Rio de Janeiro became the next site of exploitation and corruption.[2] Drive to regenerate and redesign has affected vulnerable communities and districts, such as favelas (slums) and self-build settlements. In this process, Vila Autódromo became the world-famous site of struggle to stay put.

Vila's struggle against disappearance was simultaneous with the fast regeneration of the docks area, where the appearance of new buildings and developments became the inaugural site of publicity of the Olympics. From the beginning, art and cultural production played an important role in the resistance, and as well it was also set to design profit via new developments and regeneration projects. In the dockland area in Rio, regeneration projects included: a new art museum, a science museum, historical buildings converted to studios and offices for architects and designers and more.[3] The partnership was between the mayor of Rio de Janeiro (Eduardo Paes, from the center-right party), the state of Rio de Janeiro and the president Dilma Roussef with large private construction companies, launching the private-public partnership (PPP). The entry of a new speculative (and corrupted) economy has always depended on the capacity of the locals to adapt to a new fake aesthetics: the aesthetics of regeneration. Nevertheless, many

---

1   A total of 75 thousand people were evicted in the city of Rio de Janeiro. A chunky source of research is the dossier 'Megaeventos e Violações dos Direitos Humanos no Rio de Janeiro', published in November 2015. It was put together by the Comitê Popular da Copa e Olimpíada, a popular committee formed by housing associations and NGO's. 'Megaeventos e Violações dos Direitos Humanos no Rio de Janeiro' 13 March 2019, https://br.boell.org/pt-br/2015/12/10/dossie-rio-olimpiadas-2016-os-jogos-da-exclusao.

2   We suggest articles from Adriano Belisário to follow this thread, such as 'Lava Jato document suggests cartel created for the Olympics'. Belisário shows how companies that would win the competition for building Olympic sites were already guaranteed one year before. They are Construtora Norberto Odebrecht, Andrade Gutierrez (AG), Carioca Engenharia and Carvalho Hosken (CH) – the latter is the company that most profits from building in the Barra da Tijuca and Jacarepaguá area, where Vila Autódromo is located. Adriano Belisário, 'Lava jato sugere cartel da olimpíada', A Pública, 29 April 2016 https://apublica.org/2016/04/documento-da-lava-jato-sugere-cartel-na-olimpiada/.

3   Besides Belisário (see note above) Barbara Szaniecki, Laura Burocco, Sérgio Martins and some text from myself Cristina Ribas, develop analysis about gentrification and regeneration in Rio de Janeiro. Artist's works dealing with this subject, amongst others, are from Raphi Soifer and Guerreiro do Divino Amor.

of these projects proved to be unsustainable and failed.[4] In parallel, those who disagreed with this process have been searching for ways to practice their resistance.[5]

When becoming artists in collaboration with Vila Autódromo in 2017, we have experienced a complex situation in which we could perform a series of questionings, facing contradictions of working in between art and regeneration, cultural practice and resistance against neoliberal destruction of communities and modes of living. As said in literature we follow, 'complexity forces us to push the limits of our usual methodological procedures',[6] and yes, here we have found conditions that forced us to push and invent, in which it was inevitable to work artistically without at the same time engaging in alliances, values and images of the site and the people we were invited to collaborate with. As we develop further, the invitation was to help replicate the memory of the struggle, in creating a tool that could spread out the narratives and the value of that struggle. For us, that seems possible since we seek, as artists, to nurture political intimacy with the militant investigation.[7] Vila Autódromo community although, stressing out the idea that only a bourgeoisie produce art or art programmed to be valued by the market, created conditions for cultural production and resistance in a situation of complete abandonment by the state. Set around the lake of Jacarepaguá, Vila was occupied in the early 40s by families of fishermen (and a very known fisherwoman called Mariza do Amor Divino). The settlement grew with the arrival of builders who constructed the Autodrómo (racetrack) itself, which then gave name to the community. During the 1960s, new housing for middle and up class people was built in the area of Barra da Tijuca, but no housing for the construction workers has ever been build[8]. Vila's occupation became the home site of many of these workers, and in the 1990s, it was made legal with a concession of ownership of the land given by the governor of the state of Rio. In recent years, in the context of the mega events, that concession was reversed due to an agreement made by the state and the mayor of Rio, easing the negotiation with building companies who wanted to profit the mega event. Vila is the closest neighborhood to the Olympic site, situated in an area of growing land grabbing and speculation. Because of that, between 2012 and 2015, due to the proximity to the Olympic site, Vila Autódromo was almost entirely evicted.

---

4   One example is the large contemporary art market (Art Rio fair) that was settled in the docks area (in warehouses from the nearly abandoned docks site, due to a transference of the docks to a site directly on the sea). After the three editions, which saw decay in public and selling, the Art Rio fair moved *back* to the south of the city, settling itself in the Marina da Glória.

5   Among them: Armazém Utopia collective that has produced theatre and dance plays, as well as those gathered around Carnival warehouses that became a flame of resistance in struggles for the right to the city; struggles against racism and against the commodification of culture.

6   Eduardo Passos and Regina Benevides de Barros, 'A cartografia como método de pesquisa-intervenção' in Edvardo Passos, Liliana Escossia & Virginia Kastrup, *Pistas para o método da cartografia*. Porto Alegre: Sulina, 2009, p. 30.

7   We don't presume that the connections between art and militant investigation are set; we say that relating broadly to the field of knowledge and practices that persist outside academia and on its borders, as mapped out by Timo Bartholl in *Por uma geografia em movimento: a ciência como ferramenta de luta*, Rio de Janeiro: Consequência, 2018.

8   See artist's work *Paraíso Ocupado* (Occupied Paradise) by Wouter Osterholt and Elke Uitentuis to better understand Barra building companies economies and Brazilian modernism. Wouter Osterholt & Elke Uitentuis, 15 April 2019, http://www.wouterosterholt.com/paraiso-ocupado/paraiso-ocupado-3.

Fig. 1: View from Vila Autódromo to the Jacarepaguá Lake. The beautiful landscape is one of the main reasons for the interest of real estate speculation in the territory. Photo: Cristina Ribas.

When we speak about Vila Autódromo's struggle today, we are referring to a group of no more than 15 families that remained in the area, where around 2000 people used to live. Many leaders from the residents' associations were on the front line of this organized struggle (such as Inalva Mendes Brito, Altair Antunes Guimarães, and others). Most of them gave up out of exhaustion in the harsh months of the eviction. There is no precise census of the people living in the area. Nevertheless, we know that there were around 650 families organizing against eviction. They used to live in 650 houses – in very different conditions, from barracos (improvised, poor and precarious homes) to three stories houses with a garage. The dismantling of the Vila has been initiated simultaneously with the eviction of favelas. The mayor of Rio de Janeiro in 2009 classified favelas as 'risk areas' due to the landslide after excessive rain. This has coincided with Rio becoming a site of the Olympics. The back and forth of negotiation between the housing department and the inhabitants of several communities to be evicted lasted for years. Communities wanted to negotiate in a group, trying to preserve the bond between the families and the relation to the territory, not to succumb to individual agreements as forced by the city. After being almost completely demobilized, the political energy never left Vila Autódromo. Vila was to remain. A re-urbanization plan was approved in agreement with the mayor of Rio de Janeiro, and it included building 20 new homes for the remaining residents of Vila.

The homes were ready just a few days before the opening of the Olympics in 2016. Promised school and a health center, were never completed.[9]

In 2017, we were invited to engage with the *Céu Aberto project*, part of the larger project by Goethe Institute *Future of Memory*. The purpose of the project was to map initiatives that have been addressing political memory in Latin America. Initiators of the project made an attempt to connect the memory of the struggle with the cultural production and narratives about human rights. Since the beginning of 2017, several art projects have been developed in the frame of this project.[10] We have been invited by Gleyce Kelly Heitor, a museologist and a curator. Gleyce urged us to use a methodology similar to the one that we have developed in the project *Political Vocabulary for Aesthetic Processes*,[11] and apply it to the struggles in Vila Autódromo. The goal was to create a glossary about the memory of the struggle but also opening up space for the analysis of the forces at play. How to differ from the several other collaborations and art works, previous and ongoing with the mega events, without falling on the trap of art being captured by the creative capital value that overlaps and codifies art and its events? In the observation of the conditions set by a global economy, as André Mesquita analyses, we held to the local forms of cultural production as ways to add up our artistic and militant procedure, a glossary-from-art and from listening.

On a larger scale, the reciprocal permeation between culture and capital is reproduced in international commercial trades, in intellectual property laws, in control on immaterial labor, and in the circulation of the axes of the creative sector in the global cities, from tourism bound to art through an investors class which legitimate their status by sponsoring multinational museums (as Guggenheim), or in the process of urban regeneration and gentrification, whose investments in cultural spaces in cities downtowns and the instrumentalized use of subcultural aesthetics reinforce the profit and prestige of this structures. Within this frame, the resistance of an opposition culture might fall into hypocrisy.[12]

In Vila's struggle, this meant recognizing and coming together with the practices of resistance performed by Vila itself, such as the Museu das Remoções and a large archive of images organized by Luiz Claudio da Silva, *Imagens de Memória e Luta* (Images of Memory and Struggle).[13] Memorialization of Vila's history comes with Luiz Claudio's archive and gains another

9   The Popular Plan to regenerate Vila merging with the Olympic site was never considered in the approval of the plan made in agreement with the city mayor. 21 April 2019, https://vivaavilaautodromo.wordpress.com/plano-popular-da-vila-autodromo/.
10   Curators were João Paulo Quintella, Igor Vidor, Shana Santos e Gleyce Kelly. *Futuro da Memória*, 25 April 2019, https://www.goethe.de/ins/co/es/kul/sup/mem/ciu/rio.html.
11   The book and project *Vocabulário político para processos estéticos* was conceived by Cristina Ribas. She invited around 20 people to join in a collective investigation of passages between aesthetics in politics, also considering the cycle of protests in Brazil since 2013. Cristina Ribas, 'Vocabulário Político para Processos Estéticos', 19 April 2019, https://vocabpol.cristinaribas.org/ , small version in https://desarquivo.org/node/31643.
12   André Mesquita, *Insurgências poéticas: arte ativista e ação coletiva,* São Paulo: Annablume, Fapesp, 2011, p. 234.
13   Luiz Claudio da Silva, *Imagens de Memória e Luta* (Images of Memory and Struggle), 20 June 2020, https://www.facebook.com/imagensdememoriaeluta.

collective agency in 2016 with the Museum of Evictions (Museu das Remoções)[14]. This open-air museum brings forth the memory of the struggles that took place in the community area. The Museum of Evictions is an instrument for resistance and struggle, with national importance, reaching out to communities that suffer or have already suffered from evictions and speculative practices.

Our goal is to fight eviction policies, arbitrary actions of their initiators and the consequent erasure of the memory.[15]

Fig. 2: Maria da Penha Macena and Luiz Claudio da Silva in Araçatuba São Paulo, where they went to participate in an event to support a community to be evicted. Photo: Archive of Imagens de Memória e Luta, 2019.

The Céu Aberto project learned from the Museum and wanted to join its efforts. The Museum created landmarks to help the tenants relocate since the area of the Vila was almost completely washed away in three years previous to the Olympics. There are very few traces from the previous life, such as floors of houses and tiny parts of walls. Streets, houses, shops and common areas were totally cleared out slowly, spreading debris and transforming the area in an inhospitable site due to the constant work of bulldozers and moving walls which slowly were demarcating the gaining of land for the Olympic site. At the site, a gigantic parking lot for the guests and visitors of the Olympic's was build. Today at the same location, gigantic commercial music festivals are taking place.

---

14   The museum was idealised by Thainã de Medeiros, with the support of Diana Bogado, Mario Chagas and other museology students. Museum's website: Museu das Remoções, 18 April 2020, https://museudasremocoes.com.

15   *Museu das Remoções*, 18 April 2020.

Fig. 3. Participants of one of the workshops of Céu Aberto project walk in the area of Vila Autódromo and meet the luxury hotel built for the 2016 Olympics, which became a division line between the community and the Olympic site. Photo: João Paulo Quintella, archive of the project Futuro da Memória.

In 2017, the Museum of Evictions donated rubble to the National Historical Museum. As Sandra Maria de Souza writes in the glossary:

> The rubble is certainly the main exhibit in our collection. Debris has been incorporated into the collection of the National Historical Museum, now on display in the permanent exhibition about contemporary history. For us, this is undoubtedly a great achievement. While fighting for the right to stay put, we ended up conquering spaces that we never imagined before.[16]

During this process, working with the history of the struggles in Vila, we have learned that the persistence of this struggle has potential that goes beyond the struggle for housing and the reconstruction of Vila Autódromo. Vila Autódromo's resistance teaches residents of Rio de Janeiro, not just those of occupied territories and favelas, about forms of collectivity and neoliberal oppression. Vila confronts modes of life and social reproduction more associated with constant privatization and alienation brought about by the alignment between market and state. This is made explicit in the landscape itself and in the name of the *condomínios* (tower blocks): right beside Vila gigantic tower blocks reproduce an ostentatious bourgeoisie living known forms of alienated life. Developments signed by Carvalho Hosken company reinforce that with names such as *Pure Island*.

---

16   Cristina Ribas and Lucas Sargentelli (eds), *Vocabulários em movimento,* Rio de Janeiro, 2017, pp. 22

Cristina went to Vila Autódromo for the first time in 2009 with other participants of the Universidade Nômade network, public prosecutors and representatives of several movements and activist networks,[17] that worked in different districts to map out evictions. After the invitation to make the glossary in 2017, we (Cristina and Lucas) have developed other collaborations and artworks about the history of Vila.[18] In this essay, we would like to share our relationship to the history of the Vila Autódromo struggle from three perspectives. The first one is our experience with creating the glossary *Vocabulários em movimento ⋀ vidas em resistência* (Vocabularies in movement ⋀ life's in resistance)[19] together with nine residents of Vila and the project curators. The second perspective wants to shed light on the role of women and the role of care in the struggle. In the final part, we will reflect on the role of cultural production and theatre practice in housing struggles by reflecting on our own experience.

Fig. 4 and 5: Posters (2019) developed with photos from the Imagens de Memória e Luta archive (Luiz Cláudio da Silva) and photographs from us Cristina Ribas and Lucas Icó.

---

17   Two *núcleos* (squad) were the most active in the *Defensoria Pública* (Public Defenders): Núcleo de Terras e Habitações (Nuth) and Núcleo de Defesa dos Direitos Humanos (Nudedh), which were dismantled in the next years. Alexandre Mendes, 'A nova luta da Vila Autódromo e dos moradores que resistem à remoção: reconstruir a Defensoria Pública e sua autonomia', *Universidade Nômade*, 16 June 2021, https://tinyurl.com/s5va7lo.
18   In 2019, we went back to Vila several times with the filmmaker Sol Archer. We produced three videos and three posters to be shown in an exhibition curated by Adeline Lépine in Montbéliard, France. This time we focused on the walks of Denise (one of the residents), theatre practices, and the archive of photographs of the eviction process and the cultural occupations.
19   *Vocabulários em movimento ⋀ vidas em resistência*, in portuguese, 16 June 2021, http://desarquivo.org/node/31646; and in french https://tinyurl.com/2wntntuc.

Vocabularies in movement

The eviction of the Vila intensified around 2013 and coincided with the Revolutionary Spring that came after the cycle of urban struggles in 2011. The 2013 June uprisings was a peak of the resistance against the continuing precarization of life during the *developmentist* governance of Dilma.[20] In 2017 when we arrived in Vila, there was a overwhelming general sense of suffering and feelings of loss. Nevertheless, the narratives of the people who remained in Vila have been the ones of strength, determination and astonishment at their own victories. The Macena family (Maria da Penha Macena, Luiz Claudio Macena and their daughter Natalia), as well as Sandra Maria de Souza, Sandra Regina, Jade Sol de Souza Teixeira, Denise Costa dos Santos, Dalva Chrispino de Oliveira), participated in the making of the glossary.

Getting in touch with these narratives through the project and producing the vocabulary has been a delicate operation. First, we had to come to the understanding of what forms of art could be used in a context that was already using art to seal big contracts and hide oppressions.[21] Working like artists, learning from other practices such as ethnography, cartography, the social clinic, and from our other experiences, we wanted to understand, above all, how we could be useful to the people of Vila; how we could add to the material and affective needs of the people in struggle.[22]

The history of Vila Autódromo's resistance has been going on for more than five decades. Our cartography practice revealed multiple points of view and revealed the connection between resisting and reproducing Vila's community. This is how Nathalia and Sandra residents of Vila and participants in the project, described their experience:

> After the meeting of the Residents' Association we used to walk through the streets of Vila. Their task was to collect signatures for the petition against the eviction from residents. Door knocking and walks in the neighborhood, allowed them to get to know residents faces, their lives, and routines better. Some of them they have already known. Some of them more, others less.[23]

We started with cartography as a form of research, a type of cartography always partial, never total.[24] It is a non-representational device, a way of getting to know, and perhaps a way of producing art. It helped us designing a temporal approach, the singular encounters to perform listening as a device to get to know and to engage. Cartography as a research (anti)method has

---

20   The Swiss film from Samuel Chalard, *Favela Olímpica* (2017), shows the different spaces – where negotiation with investors was happening and the reality of evicted areas.

21   When MAR – Museu de Arte do Rio was open in 2013, there was massive protests outside, even though the curators and the public program tried to deal with past conflicts from other contexts – like inviting groups from São Paulo and MTST movement to present some of their collaboration from beginning of 2000.

22   Lucas, who also works as a graphic designer, redesigned together with Luiz Claudio Vila Autódromo's entrance sign, which exhibits a photo of before and after.

23   Cristina Ribas and Lucas Sargentelli (eds), *Vocabulários em movimento,* Rio de Janeiro, 2017, p. 13.

24   Suely Rolnik, *Cartografia Sentimental: Transformações Contemporâneas do Desejo*. Porto Alegre: Sulina/UFRGS, 2011.

enabled us to get to know people and their stories, to talk about the struggle, to create modes of expression, and trace the inseparability of these acts from one another.[25]

Our initial proposal was simple. We proposed to have conversations with the residents, to get to know them, and to learn about their relationship to the struggle. We asked them to bring at our encounters three *strange* words, an object and a sound. We met in one of the new homes that was built in 2015/2016. Between its empty walls, we created together a temporary reception for visitors of Vila. After mapping the conversations, we created a map of themes relying on conversations with nine residents and our own research about the struggle. We made notes, drew diagrams and discussed with residents. Curators of the project were also invited to contribute. The idea was for everyone to write from a personal perspective but in connection to human rights and politics of memory.

As a result, the 44 pages publication was printed in Risograph and distributed by the Goethe Institute free of charge. The texts in the publication are very different in nature. There is a text about the Museu das Remoções, public utility information, short entries about the context of the Olympics related evictions, fragments of conversations presented as *wavy words* (phrases in the shape of waves), prospective diagrams and diagrams showing the structure, the history and the future of Vila.

We can think about the glossary in light of Martha Rosler's experience:

> Consider the city once again. It is more than a set of relationships and a congeries of buildings; it is even more than a geopolitical locale – it is a set of unfolding historical processes. In short, a city embodies and enacts a history. In representing the city, in producing counterrepresentations, the specificity of a locale and its histories becomes critical. Documentary, rethought and redeployed, provides an essential tool, though certainly not the only one.[26]

The publication came out as a way to reproduce the history of resistance of Vila Autódromo, but it also contains an explicit preoccupation in sharing the method that we have created. As artists, we became good listeners and maybe *translators* (or transducers) of the embodied memory of the struggle. We asked ourselves: How do we deal with this intimacy that opened up between us? How does it produce expressive modes? How do memories and voices become content for the glossary? How are the voices and the memories of the residents spoken now from our bodies? How does the glossary continue to mobilize vocabularies from struggles?

We decided to call the publication *vocabularies in movement, lives in resistance* to reinforce the

---

25   The purpose of cartographies is *to evidence a plan the forces*. When we talk about cartography as a method for research, we say it from what was developed in Brazil, upon Gilles Deleuze and Félix Guattari's theories and practices, philosophy of difference, group devices and institutional analysis. Liliana da Escóssia, Virginia Kastrup and Eduardo Passos, *Pistas do método da cartografia. Research-intervention and production of subjectivity.* Porto Alegre: Sulina, 2009.

26   Martha Rosler, 'Fragments of a metropolitan viewpoint', in Brian Wallis (ed.) *If You Lived Here. The City in Art, Theory and Social Activism. A project by Martha Rosler,* Seattle: Bay Press, 1991. p. 32.

continuation of this struggle, the lives of the people. Recalling it now, we did this with the idea to actualize this unfinished struggle in a still very segregated area of Rio de Janeiro, but that does not move back in the staying put of adding another layer in its complicated history. We realized that speaking and listening configured another form of approach to struggle, moving beyond forms of academic research and textual production about social movements, tensioning with the militantism in our bodies and the possibilities of art.

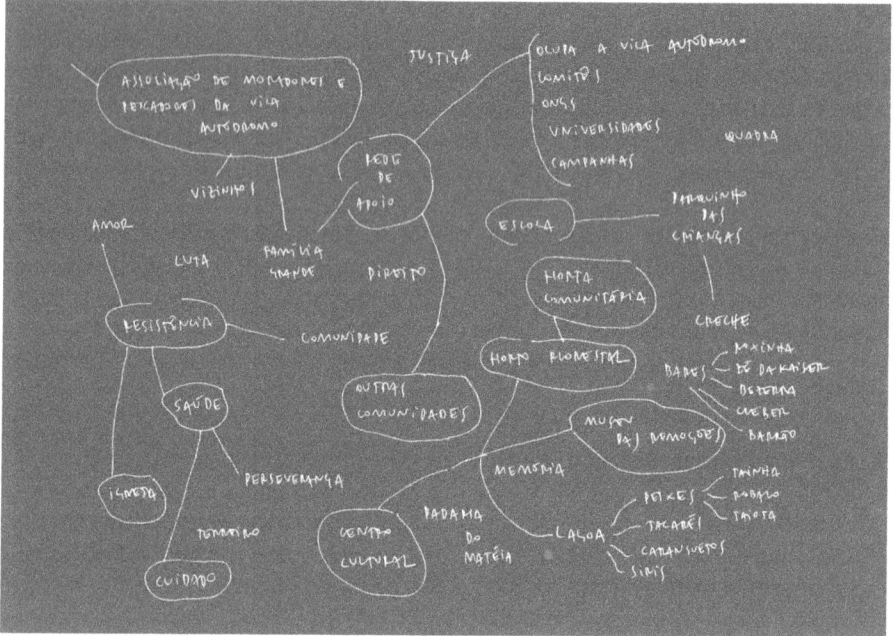

Fig. 6: Cristina's diagram from conversations with residents, relating past and future: what was there and what is planned to be rebuilt, 2017.

The role of women and care networks

Mariza do Amor Divino told us in the conversations: 'fear is the weapon of the weak. My grandma always told me'.[27]

One of the most important aspects of Vila's residents' struggle, has been the distinguished role of women in the organization of the resistance movement. Many households were managed solely by women. One of our aims has been to highlight their role in the context of ethics of care, which in Vila relates to the production of a close relationship to the environment and maintenance of community life itself. The difficulty here is how not to reproduce a naturalized relationship between gender and care while talking about women's role in defending their communities, despite precarization of their lives.[28]

27   Cristina Ribas and Lucas Sargentelli (eds), *Vocabulários em movimento,* Rio de Janeiro, 2017, p. 44.
28   Cristina Ribas, 'Care Denial: Reproductive Care and Care as a Right', *Revista Mesa,* 2018, *20 May 2021,*

The protagonism of women residents is directly connected to the fostering of care networks and to the maintenance of infrastructure to guarantee the daily life of the community, previously and through the eviction. 'Struggle was made with many hands'. This is how Maria da Penha describes life in resistance to eviction. During the eviction years, the community had to face violent pressures to move out, including water restrictions, police repression and psychological pressures. 'Strength was in collaboration.'[29] One of the critical scenes of everyday life was the commercialization of food for Vila residents' and for the demolition workers These people had to face similar situations in their own districts.

Invisible domestic labor characterizes women's everyday life. The very production of the feminine gender is historically directly associated with their homes and to care work (either reproductive work, care of children, elders, or neighbors). In this process, we have learned that women in Vila persistently resisted the violent destruction of care infrastructure. This is why we have included entries such as: 'Women's role', 'Women who mend pipes/Feminist economy', and 'Support network' in the publication.[30]

A number of researchers have focused on the role of women in anti-eviction struggles in Vila. In her book *Human rights and the colonization of the urban space* (Direitos Humanos e a Colonização do Urbano) Marcela Munch has been focusing on the history of Vila and the resistance from women's perspective.[31] In the conversation with Sandra Maria, one of Vila's tenants, we have learned that she doesn't identify as a feminist, but she recognizes the specific role of women in struggle. She has told us that men have to follow women, support them because they – the men – have always been in the position of privilege. Sandra Regina, who has also resisted eviction and remained in Vila, confirmed that she has been one of the main protagonists responsible for repairing water pipes supplying residents with water in the years of eviction.

The neoliberal economy is known for directly destroying communities, as said by many movements and authors from political organization, such as Jeremy Gilbert,[32] to feminisms, such as Mariarosa Dalla Costa[33]. What we also know is that the model of the individual, exploited and indebted, is not a plan of success. It is violent meritocracy, a push into people's lives and does not respect cultural singularities. Informality has also been very present in the living and working conditions of the people in Vila. In many cases, informal work (including different services) was constitutive for the community.[34] In relation to this, it made sense to approach

  http://institutomesa.org/revistamesa/edicoes/5/cristina-ribas/?lang=en.

29   Cristina Ribas and Lucas Sargentelli (eds), *Vocabulários em movimento,* Rio de Janeiro, 2017.

30   If we had more time, we could have discussed concepts that make sense here, such as the feminist economy. We were happy to get to know that Instituto PACS was developing the concept of feminist economy and communitarian economy also through cartography in close areas and communities. See *Militiva,* Instituto PACS, 23 April 2020, https://www.militiva.org.br/mapa.

31   Marcela Munch, *Direitos Humanos e a colonização do urbano.* Rio de Janeiro: Ed. Lúmen Juris, 2017.

32   Jeremy Gilbert, *Common Ground: Democracy and Collectivity in an Age of Individualism*, New York: Pluto Press, 2014.

33   Mariarosa Dalla Costa, *Women and the Subversion of the Community* [1971], in Barbagallo Camille (ed) *Women and the Subversion of the Community: A Mariarosa Dalla Costa reader.* California: PM Press, 2019.

34   Ricardo Nery Falbo, 'A Comunidade Vila Autódromo na fala de seus moradores: um relato atípico de

the conversation starting from questions: how does the struggle interfere in everyday life? When is the status of *work* dislocated? How does struggling oppression take over normal flows of life, often making it impossible to work? Unluckily, as we have heard from those involved in the struggle, it could happen that even though you made sacrifices and earned less while working for the movement, at the end, 'you could also lose the place you live'.[35]

Care has been building the fabric of the collective structures in Vila Autódromo. Nevertheless, community has been broken many times. Because the movement has been working on the collective legal defense together with the public prosecutors, the mayor of Rio made a lot of psychological pressure on tenants in order to dismantle the community in the struggle to be able to deal with the cases on an individual basis and convince them to take the new contracts. For tenants talking about this is still bringing uncomfortable feelings. This is one of the ways that the community got broken. Since then, it has been recovering step by step. Nowadays, the church works as a gathering space for the community. Some of those that had left Vila often come back to take part in the prayers – as well as some middle-class neighbors from the nearby tower blocks.

Fig. 7: Lucas walking with Denise, frame from the video Caminhar para longe (Walk afar). Photo: Sol Archer.

While interviewing the tenants for the glossary, we found out that Denise developed a walking practice as a form of self-care. With great vigor, Denise has been walking up to three hours a day to keep her body and mental health in a good shape. In 2018, in collaboration with Denise and film maker Sol Archer, Lucas has made a film.[36] Departing from Vila, Denise's walks reach out to other places important for her-story. The walk has also been an orientation challenge for her. Many of the residents reported the feeling of disorientation that the erasure of the Vila has

---

fatos, temas e questões sociopolíticos' in Alexandre Mendes, Ricardo N. Falbo and Ricardo Teixeira, *O Fim das Narrativas progressistas na América do Sul*. Juiz de Fora: Editar Editora Associada Ltda, 2016. pp. 137-154.

35   Ribas and Sargentelli (eds), *Vocabulários em movimento*, pp. 42.

36   Lucas Sargentelli Icó, 'Uma caminhada com Denise', *Fotocronografias,* 25 December 2019, https://medium.com/fotocronografias/uma-caminhada-com-denise-255a09e4a5e1

created. To resolve this problem, they have created orientation points with placards that signal where houses, streets and public places use to be. These signs became a part of the open-air Museu das Remoções.

The relations based on care in Vila extend beyond inter-human interaction to an ecological relationship with the site itself. After the destruction of Vila tenants have been replanting trees, to re-establish an intimate connection with the surrounding life. To celebrate these relations, we have introduced in the glossary the entry called 'Trees'. In this entry, we have listed the types of trees that exist in the area. The mayor ordered the cut of more than 400 trees in the eviction process.

In the glossary, we have wrote: 'In this place, there is a co-extension between the care for oneself, the care for the family, the neighbors, the surroundings and nature. Care gives tissue to the collective structures, erased by the neoliberal mechanism of evictions.'

Theatre practice and cultural production

Making the glossary with the tenants was also about finding out more about their personal lives. When we talked to Natalia Macena in the making of the glossary, we found out that she studied theatre in the university. She also told us that she used to make theatre in Vila itself! The conversation about her experience of eviction and how they used to organize in Vila, brought another rhizomatic line to this complex history. Since the mid-1990s, her father, Luiz Claudio used to organize theatre performances for the annual festivals. They were organized as a benefit to building the Catholic church – São José Operário (Joseph, the Worker), the only building that remains on-site – undemolished from before. Luiz Claudio has been responsible for the scripts that have been written based on the biblical stories or fiction taking place in the favela. Many of these scripts talked about the class struggle between the poor and the rich. We invited Natalia and Luiz Claudio to write about this for the glossary. They have written about the role of the theatre and the struggle.

One year later, at the end of 2018, we went back to Vila with the idea to use theatre tools to revisit memories about the struggle. It seemed to us that through the embodiment of theatre, body expression and improvisation, we could create the conditions to talk about the past. We explained to the residents that we wanted recall the past that is not based only on talking and seeing the photos. Our idea was to go together with the current residents that took part in the project to the places where old residents have been moved to (Parque Carioca, a poorly state housing development[37]). We wanted to do interviews and to use theatre games to remember parts of the plays performed in the 1990s to relate to the present. We really wanted to hear the voices of those who have left. Nevertheless, residents that remained in Vila didn't feel

37   Residents moved to many other places, spread, making it harder to get together. We don't know a map of displacement of Vila's residents yet. The militia is a form of militarised power that controls areas in Rio de Janeiro. The control aims at bringing safety (which means not allowing drug dealing to happen in those areas), charges money for services such as gas, water, internet, and often controls forms of public expressions, such as cultural events. Usually a mixed constitution of police and reformed army.

comfortable visiting them. For them it felt like a painful future they could still avoid. But the new buildings were taken by a third power (militia), where it would not be possible to film or walk freely. Despite this, the residents that remained encouraged us to continue. They chose to organize the encounters in the Vila itself. It made a lot of sense to us. After all they were the protagonists of the project!

In one of the photos that tenants brought to our workshop showing the Festival of the 1990s we recognized Maria da Penha and Luiz Claudio current residents of Vila. They both used to do theatre when they were young, living in other communities. They took part in projects in different contexts: in Rocinha favela, Maria da Penha when she was a kid, and Luiz Claudio when studying at the physical education faculty. As a starting point of our process, Luiz Claudio organized an exhibition of the photos from his photo archive *Imagens de Memória e Luta* in the church. There we heard more about the plays, saw videos and the original scripts. We decided to reenact several scenes by using theatre tools. This process has taught us about the past, but also about the present. The process was also about sharing emotions, liveliness and joy. Merging the borders between life and theatrical play, Natalia, for example, had recalled how her mother was dancing and celebrating when she took the trash out after she got the confirmation they could remain in the middle of the destroyed Vila in 2015.

We have been working in the past years with bodily and group dynamics in the context of the arts, and Theatre of the Oppressed, the method devised by Augusto Boal. The bodily and improvisation games developed in the Theatre of the Oppressed want to bring attention to the class dynamics. Here, the theatre games could also be used to remember the everyday of evictions. We could have re-enacted a few critical moments to work out more literally the dynamics of the oppression, an analysis of the oppression and the resistance of the oppressed.[38] This, in the end, seemed quite unnecessary. We felt a need for celebratory moment now that the hard years of eviction have passed. There was a need to come together, to (re)invent connections, to tightening love and care bonds again. For our workshop, some people that had not returned to the Vila since the eviction came back for the first time. This was very moving. Many children took part in theatre plays in the 1990s. They were also invited to join the workshop – except now as adults, and bringing their kids to play and learn from the group experience. Cristina´s daughter, Hannah, was there too.

Cultural production was a way of gathering support in the eviction years. In 2013 tenants of Vila have organized with many collectives such as the Ocupa Vila Autódromo (engaged in the Occupy movement). They used music, cinema, theatre and more. As artists we had our own way of encountering. Discovering the struggle and its connection to culture and art was like a gift to us. This process has re-located our own creative approach, becoming listeners also of cultural production in the making of resistance and the community. Politics and art, for us, also an ethico-aesthetics could be reaffirmed as in Vila Autódromo, that art practice is a form of encounter and expression of life itself.

---

38   Several methods of Augusto Boal are applied by us (such as Image Theatre and Rainbow of Desires), but we do asset that we do *unorthodox* Theatre of the oppressed. Augusto Boal, *Theatre of the Oppressed*. London: Pluto Press, 2008.

In her text for the vocabulary, Natalia wrote:

> *In the struggle, it was necessary to reinvent, resist, re-exist! Between hammering,*
> *blows, tractors, aggressions, rubble and a lot of fallacy ... a supernatural force*
> *(instinctively collective) emerged for survival in the adored land, and hammered in*
> *mind: if we join our forces with our creativity, we can reverse this situation! But action*
> *is needed! And if there is something I learned in this eviction process, it is that we need*
> *to leave the role of mere spectators to be actors of our own history. (...) May we have*
> *wisdom, ardor, courage and faith to navigate this sea of land, that we can act in both*
> *spheres:in Life and in Art!*[39]

Fig. 8 and 9: Presentation of footage to the group, 2019 (top). Theatrical improvisation workshop, 2019 (bottom). Photos: Sol Archer. .

39   Ribas and Sargentelli (eds), *Vocabulários em movimento*, p. 6.

Where do we stand?

Imbrications between aesthetics and politics in the context of mega-events reinforce and generate various forms of exclusion in the city of Rio de Janeiro, as well as in Brazil. Art needs to criticize this subsumtion, since it is made an instrument of regeneration and gentrification, diminishing and erasing class and territorial struggles. At the end of the Céu Aberto project, there was the possibility to organize a public event in the MAR museum, in the docks area. There was a common understanding amongst people in the project Céu Aberto, that the city regeneration project that built MAR Museum was the same plan that wanted to evict Vila. Therefore, it didn't make sense to organize the closing event in the Museum. Calling the artistic community to travel to Vila for the final event, instead, was an attempt to break the privilege of certain areas by giving agency to community cultural production and bring the cultural class closer to struggles. Our role as critical artists is to prevent the institutions and art markets from profiting from this material without any gain for the community. We situate ourselves in solidarity with the struggle, always in conflict with the project of commodification of the city and its forms of life. This also means producing other forms of discursive positioning(s) in the artistic context. This is a principle for us. We are dedicated to the struggle, not just by engaging with the memory of Vila Autódromo against its disappearance, but also by continuing struggle against the commodification of cities and lives in other work that we do.

To stay within the movement, the videos and the artworks, as well as the produced knowledge, must stay connected to the struggle. Everything that has been produced about the struggle has been shared with tenants of Vila including the income that has been made. From intensive listening, we have learned how to compose with/from the local conditions, desires and material needs. These works we have developed, as a form of the aesthetic invention, inevitably resituate us in relation to our own communities (this means those effects that should also be taken or assumed by the militant researcher, the effects of the research in the researcher itself). The struggle against the disappearance of Vila should be then not just the right to remain physically in the territory and the right to keep your home, but also the right to sustain the relationships and the values that structure that mode of living – culture itself.

As Marcela Munch, a local researcher said to us: 'the struggle expresses a force of resistance to the modes of commodification in contemporary cities, and it has an unusual persistence in the fabrication of the city.'[40] This struggle is not only about staying put but about new forms of collectivity emerging under constant privatization, alienation, suffering and indebtment, hence, the right to preserve the bonds that are constantly erased by capital.

---

40   Marcela Munch, *Direitos Humanos e a colonização do urbano*. Rio de Janeiro: Ed. Lúmen Juris, 2017.

References

Bartholl, Timo. *Por uma geografia em movimento: a ciência como ferramenta de luta*, Rio de Janeiro: Consequência, 2018.

Boal, Augusto. *Theatre of the Oppressed*. London: Pluto Press, 2008.

Bogado, Diana. 'Museu das Remoções da Vila Autódromo: Resistência criativa à construção da cidade neoliberal' In: *Cadernos de Sociomuseologia* No. 10, (2017).

Dalla Costa, Mariarosa. 'Women and the Subversion of the Community' [1971], in Barbagallo Camille (ed) *Women and the Subversion of the Community: A Mariarosa Dalla Costa reader*. California: PM Press, 2019.

Escóssia, L., Kastrup, V. & Passos, E. *Pistas do método da cartografia. Research-intervention and production of subjectivity*. Porto Alegre: Sulina, 2009.

Falbo, Ricardo Nery, 'A Comunidade Vila Autódromo na fala de seus moradores: um relato atípico de fatos, temas e questões sociopolíticos' in Alexandre Mendes, Ricardo N. Falbo and Ricardo Teixeira, *O Fim das Narrativas progressistas na América do Sul*. Juiz de Fora: Editar Editora Associada Ltda, 2016, pp. 137-154.

Gilbert, Jeremy, *Common Ground: Democracy and Collectivity in an Age of Individualism*, New York: Pluto Press, 2014.

Mesquita, André. *Insurgências poéticas: arte ativista e ação coletiva*, São Paulo: Annablume, Fapesp, 2011.

Munch, Marcela. *Direitos Humanos e a colonização do urbano*. Rio de Janeiro: Ed. Lúmen Juris, 2017.

Ribas, Cristina T. 'Care Denial: Reproductive Care and Care as a Right', *Revista Mesa*, 20 May 2018, http://institutomesa.org/revistamesa/edicoes/5/cristina-ribas/?lang=en.

Rolnik, Suely. *Cartografia Sentimental: Transformações Contemporâneas do Desejo*. Porto Alegre: Sulina/UFRGS, 2011.

Rosler, Martha, 'Fragments of a metropolitan viewpoint', in Brian Wallis (ed.) *If You Lived Here. The City in Art, Theory and Social Activism. A project by Martha Rosler*, Seattle: Bay Press, 1991.

Ribas, Cristina and Sargentelli, L. (eds), *Vocabulários em movimento ∧ vidas em resistência*. Rio de Janeiro: Edition of the artists and Goethe Institute, 2017.

SOME LOCAL NOTES ON RESISTING ART IN THE BIGGER PICTURE: POSSIBLE LESSONS FROM SOUTH LONDON

SOUTHWARK NOTES ARCHIVE GROUP

The council has been unscrupulous since the outset of the regeneration in everything they have done. This whole scheme has been a shambolic act of deception on a grand scale.

– Dylan Parfitt, ex-Heygate Estate resident

When you Find Yourself in a Dark Forest...

After all our campaigning and battling to save our neighborhood, it takes just a moment of being away from The Elephant & Castle in South London to finally understand what's maybe going to happen there. We are lucky enough to be in Milan but unlucky enough to be passing by Stefano Boeri's famous Bosco Verticale, two daft tree-covered luxury apartment blocks. Alongside this forest of capital and speculation resides an adjacent landscape of shiny corporate offices and a shopping mall from Hell that has heavily impacted the former neighborhood that is close by the Porta Garibaldi train station. What was here before, in part, was the Isola Art Center, a long-term squatted communal space focused in part on the practicalities of the years long fighting alongside residents and traders for the Isola neighborhood. Standing underneath the absurd *eco-gentrification* of the Bosco Verticale, we had a shared moment of dislocation and a poignant understanding that neighborhoods can be rapidly changed just like that when capital decides it wants to land where you live and work and be.

Back in South London the collective Southwark Notes is made up of people who somehow came together over a long period of time to work collectively to resist the ongoing disaster of a state-led *regeneration* project for the Elephant & Castle and Walworth districts in North Southwark, London. Although this means doing a ton of writing, organizing actions against the local council, anti-gentrification walks and teach-ins held in moments where the community occupied council offices and offering solidarity to other housing struggles, the work has only been able to be as strong through collaboration with local people and local campaigns. Standing beneath Boeri's monument to eco-gentrification in Milan we felt finally the full weight of what it will mean when the famous Elephant & Castle Shopping Centre, the heart of the local community, is demolished to make way for over 900+ new expensive apartments and a new demographic of both more middle-class folks but not only those. The Elephant area now has £3+ million penthouses in 44 story tower blocks when 10 years ago people would call the area a shithole and ask us if we weren't afraid to live there.

We went back to re-read the incredibly useful book *Fight-Specific Isola*, a dense and inspiring documentation of every battle both the Isola community and the Isola Art Center fought against mega-development and mega-speculation. The book is happy to talk about the mistakes they made and about those who finally jumped ship to work and be bought by various developers involved in the destruction. We had read this book before but it was only when standing in the aftermath of the destruction of part of the Isola area as detailed in the book, that we could appreciate the doom and gloom to come when the Elephant Shopping Centre gets demolished. You can call it a kind of retroactively charged poignancy because this whirlwind contains a full circle. One of us was involved in later years of the resistances around Isola and the squatted factory, first on site and then post-eviction and demolition when the Isola Art Center moved to operate as a dispersed site in the neighborhood. We don't think we had understood enough from our comrade's experiences in this struggle what it actually meant to have lived through this once but also to be facing something similar again in The Elephant.

Looking Forward to Southwark Notes...

As a group, Southwark Notes had lived through an intense involvement in the amazing campaign against the demolition of the public housing blocks of Heygate Estate. Home to 3000+ people across 1000+ secure and affordable council flats, the whole estate first had to be subject to a massive stigmatization campaign before an unholy alliance of a local Labor Party council and a giant Australian corporate property developer Lendlease got together in July 2010 to disperse the community and demolish the blocks. After this came the re-branded Elephant Park development of 3000+ private homes marketed as a new community with its private gardens and utopian communal allotments on the roofs. We had lived through our good friend's final eviction from his home, the last and only Heygate resident to be carried out of their home by bailiffs and police. Finally we had lived through our own eviction from the Heygate Estate in which we and many others had kept the place alive with film screenings, gardening, protests, music nights, tours for students and researchers, graffiti and so on. You could say that although we were not yet directly affected by the violence of a regeneration that sees thousands of working class people displaced from the homes, we are affected by the daily violence of a regeneration that takes the long-grown organic fabric and culture of our area and throws it in the bin.

As Southwark Notes working hard to name the moment, we wrote and acted on uncovering the heart of any social cleansing project namely the displacement central to such so-called regeneration schemes. Along the way of making an analysis we tried to expose the actors, the powers, the strategies being put into place to effect that displacement – phony consultation, fake viability reports from developers, the scam of *affordable housing*, revealing secret regeneration agreements, the use of stigma to legitimize chucking people out of their long-term homes. As that analysis came into being we also named the after-effects of commercial displacement and the massive increase in local private rents as a form of secondary displacement. We also used archival material to show how working class communities closer to the River Thames had been through all this before in the 1980s with the large regeneration of North Southwark under the imposed planning regime of the

London Docklands Development Corporation. Then, as now, it was common for developers to recuperate for profit any actual working class history that can be simply then be described as heritage, this being a de-politicized history solely maintained for marketing purposes. On a miserable wander through the new Elephant Park development, we read the large historical timeline of the area commissioned by Lendlease to market this new neighbourhood that features the Battle of The Pullens Estate from 1986 where tenants & squatters came together and actually saved their council estate. Just as we had been present and active on Heygate, thirty years before, some of Southwark Notes were at the Battle of The Pullens blocking the cops, breaking up the wooden panels intended to board up evicted flats and letting down tires on bailiffs vans as part of the general community carnival against eviction. It was our first ever involvement in a housing struggle in as much as the Heygate Estate would not be our last active involvement in housing struggles.

Fig. 1: Residents and their supporters on the Heygate Estate establish a community cinema to resist enclosure of the estate by developers. Photo by Southwark Notes.

Heygate Lives!

That the story of the demolition of the Heygate Estate become so well-known in London and even internationally was down, in part, to the efforts we put in over the years to make it in/ famous as both a political scandal and as a warning to other estates facing the same prospects. It has been interesting that not only have we been successful in creating a counter-narrative to the Council and developer version that seeks to justify demolition explicitly because the tenants and residents were 'the wrong kind of people' for the area, we've been adamant that we need to continue to defend that counter-narrative against any continuing attempts to dismiss it. Since 2010, Southwark Council's version of events slips randomly between guilt and no-guilt. In one moment they are saying that they would never approach estate demolition in the same way again, in another moment they attack the campaigns saying that the Council could not stand by any longer leaving their tenants in sub-standard conditions and dangerous housing. The years of managed decline and disinvestment the Council allowed on the estate sat together with a long-buried external surveyor's report that said the estate was structurally sound. For the Council the moral high ground is about as towering as a molehill.

After the demolition, when any developers or pro-development agencies came to the area, we *always* mobilized against them to stick to our belief that despite demolition, the Heygate Lives, and to hold tight to the history of the estate and the imaginaries around social housing and community that were produced in the campaign to save it. But there is also a flipside to these imaginaries when the same community decides to demand of a developer interim uses of various temporary empty sites for community benefit. Forgetting that the fight has to be *against* the developer, what comes to be accepted is a kind of received wisdom of community benefits that are then demanded from the developer. Suddenly the developer is given a human face and what arrive are dubious *developer-compliant community projects*. It was sad to see local people get involved in the *Mobile Gardeners* allotment scheme that slowly, rebranded as *Grow Elephant*, turned itself into an unpaid marketing exercise for Lendlease where its ideas of communal gardening were taking up and plastered all over the very green-colored Elephant Park hoardings. We were lacking in sympathy when one of the Grow Elephant coordinators no longer able to rent locally due to gentrification turned to living on the site in a caravan only to be evicted from this by Lendlease decree! Yes, developer compliance is a noose.

Needless to say those who had already years of doing deals with any old developer for interim-use sites for their art projects, also came to the fore. Hotel Elephant secured the old doctors surgery building on the Heygate for use as an exhibition space, for parties and for renting out as studio spaces – that old cash cow! It was no surprise then to spot Council regeneration staff in photos from parties at the Hotel Elephant space. It was all a stepping stone for a later £174,000 funding allocation from Southwark Council and the London Development Agency for a permanent Hotel Elephant in five railway arches intended to be a dedicated hub for 'emerging artists and creative entrepreneurs'. The other alleged community benefit was a Lendlease-sponsored temporary box park made up of old shipping containers sited adjacent to the Heygate demolition and full of street foods, tattoos and yoga businesses. Once again, as if Heygate never existed, some local artists and musicians saw this as a new community space to accept and appear in as performers and entertainers. It was funny how defensive or pissed off these people could be when you made a criticism of their involvement.

Pyramid Dead: Public Art as a Violence

Funny also, in 2013, was the whole saga of London-based public art commissioners ArtAngel who, working with famous artist Mike Nelson, eventually secured permission from Southwark Council to use the Heygate Estate as a site for a large 'monumental' artwork. Mike Nelson's idea was to take one of the low-rise blocks, take it apart and reassemble it as a large pyramid.

Despite some leaseholders still living on the estate and battling the Council for decent valuations for the homes to enable them to buy a new home locally, ArtAngel were keen to progress on getting the contracts signed for the massive demolition and re-construction work ahead. The Council itself was very pleased that they would be getting a temporary art piece by world-famous artist that in the logic of the unhealthy union of art and regeneration would be useful international promotion for their Elephant & Castle regeneration project.

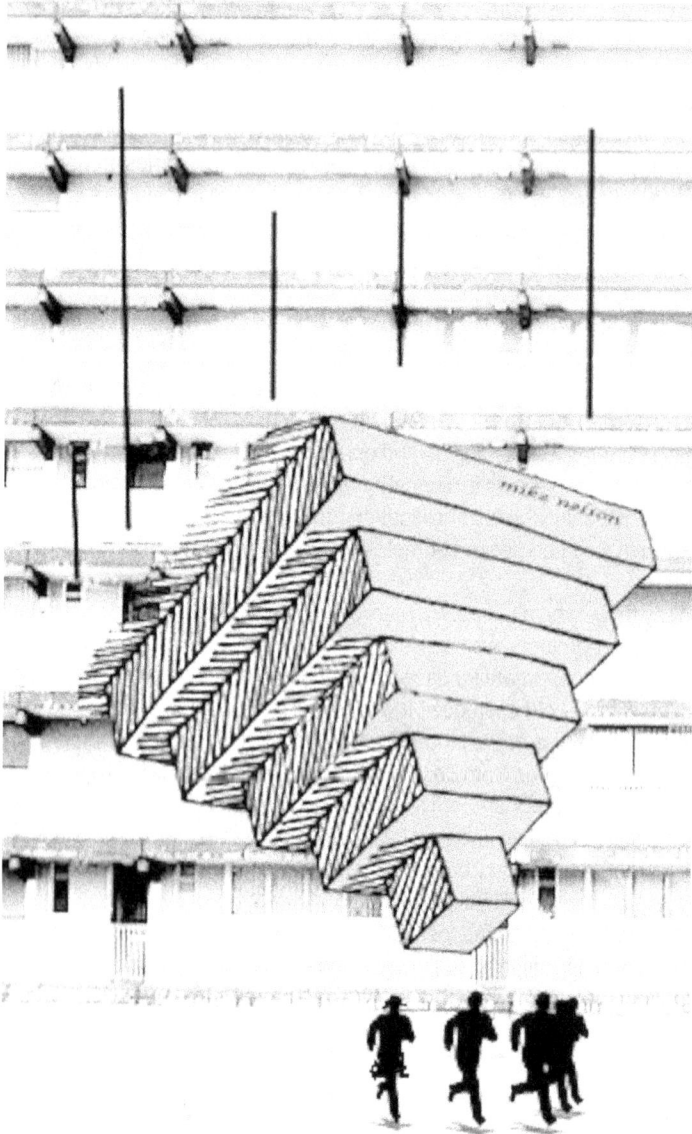

Fig. 2: Graphic design of poster that was prepared to resist the Mike Nelson artwork on Heygate Estate. Image by Southwark Notes.

When we got early wind of this project, we wrote in earnest to ArtAngel a long letter detailing why this was a bad idea. Featuring quotes from those already displaced from the estate, we argued that:

> A new and popular art work made from the material structure of the estate put on show for a publicly invited audience in a space where the space itself has been so hotly contested by residents for their homes and by those using the space for interim uses, would sound like it had been created with certain privileges that public art commissions can easily access but that local people could not. After the shoddy decant and resultant displacement of residents, the constant argument of the Council to clear the estate so that the much needed regeneration could happen, the recent fencing off of the estate denying access to numerous autonomous interim uses, we would view the then siting and invitation to an audience to view a new public artwork in this space as a gross act of symbolic violence that erases the long history and battles of local residents.

From this letter we were able to meet ArtAngel's Head of Programmes & Production Rob Bowman at a café in the Elephant Shopping Centre. Once again we argued how disrespectful and underhand the idea of public artwork would be that intended only to be a great talking point for mostly middle-class art lovers to have an afternoon out at. We said that if they wanted to hear what local people think we could put them in touch with many of the local campaigns that existed around protecting public housing and public open spaces. After the meeting, we got an email from ArtAngel saying they would love us to promote the forthcoming artwork with these groups and people! In the end what was expected to be monumental turned into a damp squib as the local opposition to the scheme was so active in condemning it that Southwark Council ended up sneaking out a Press Release in late December saying it wasn't going to happen. ArtAngel were sad but we were happy. Happy because we saw this a teaching art institutions and artists a vital lesson as well as trying to use this lesson to enable other local campaigns to learn about the revolting complicity between artists and development.

Peoples Bureau: Museumification in Progress

If the Heygate was the first domino to fall to the Council's regeneration regime, demolishing the Elephant & Castle Shopping Centre would complete Southwark's game plan. The Shopping Centre has been a vital part of the Elephant & Castle neighbourhood since the mid-60s and although much-maligned by outsiders, if you spend any time inside you would see what a beautiful monument to both working class customers and working class traders it is. People shop there but they also in some ways live little parts of their lives there too – meeting people they know, eating in cafes, hanging out doing nothing, waiting for something, hustling a bit.

Through a quite circuitous route of patronage, commissions and funding from Shopping Centre to Tate Modern to the Shopping Centre again and back to Tate Modern, the small artists group People's Bureau had been for a number of years putting on various temporary engagements in The Elephant usually based in an empty shop unit for a short period of time. This work, although under a different name goes back to 2010 and so the artists had a long-term understanding and appreciation for how people use the space and time in the Shopping Centre to do their thing. Some of us in Southwark Notes had taken part in some of these events but by the time the new owner and future developer of the Centre Delancey began to fund the People's Bureau work we were starting to feel queasy. It was always less of a question of taking the money but more what do you produce from funding that lets you create some sort of encounter between an artists group and local people? Despite the notion of empowerment for locals via differing workshops that ranged from craft activities to socials to skills exchanges, we felt that none of this really added up to making a political space where local people could produce not just good feelings about the Shopping Centre as a big part of their life but a space where local people are encouraged to act politically and communally in defense of these lives. What this means is that instead of temporary moments where local people are in some ways organically sourced at grassroots level for an art project, the art project could act, alongside the numerous local campaigns that the artists were aware of, as part of the resistance to Delancey and the long-term social cleansing of the area by the Council and development partners.

People's Bureau seemed unable to grasp the very conflict that was raging in The Elephant and merely wanted to document diversity and interesting stories as kind of time capsule for future residents of the area whoever they might end up being. Delancey had given them a trader's cart, the same type of cart that some traders trade out of in the Shopping Centre. Painted pink and acting as a repository of People's Bureau art and memorabilia, the cart would travel here and there – to Chelsea College of Art, to Tate Modern. In August 2015, we were dismayed to find this cart set up in one of Delancey's *consultation events* where the developer sought to market its plans to local people. Cheek by jowl on the cart was a mixture of People's Bureau material and Delancey's own development propaganda. Later, People's Bureau described their ambition that 'the cart will eventually return to the newly built Elephant & Castle shopping centre, thus creating the link connecting the old and the new Elephant and becoming a museum of local culture'. After an extensive and long chat with one of the Bureau where we tried to pose critical questions in an attempt to get People's Bureau to take a position on the regeneration, we thought that there maybe some shift in the way the Bureau could work. However a further Arts Council-funded *Unearthing Elephant* project was more of the same fetish for archive and documentation and a naïve notion that such material would spur Delancey to care about the traders and users of the Centre.

In late 2016, out of frustration, we published a long critical piece on the work of People's Bureau on the Southwark Notes website entitled *Empowerment for Surrender: People's Bureau, Engaged Art & the Elephant* that contained a history of their practice and a detailed account of what we thought was wrong with 'preservation into the future' the daily lives of those who inhabit the Shopping Centre in the here and now. We published a long reply by People's Bureau and a reply to their reply. It was a fruitful exchange we thought especially as we were still unclear whether People's Bureau had broken with funding from Delancey or whether Delancey simply stopped funding People's Bureau for various reasons. In 2017, as the hard work to resist the development plans coalesced with a community coalition called Up the Elephant Campaign, one of the People's Bureau artists then got involved. Using funds from a residency elsewhere in London, she initiated two editions of a newspaper called *Elephant's*

Trumpet that was written by local campaigners to inform people of the struggle at the Shopping Centre. It was a good and useful publishing project and we were happy to see something put into practice that may have come out of some of the conversations and the critical text we had written. Around the end of 2018 People's Bureau more or less ended their involvement with the Shopping Centre and also the campaigns for various reasons.

In late 2018, we had been invited to present to art students some ideas around the role of art and artists in gentrification and so we were preparing to present this story as a case study. It seemed like it could have a happy ending for once where artists do actual meaningful political work. Checking on People's Bureau website to source some quotes, we were dismayed to find them making a residency at a co-living new build housing project by property developers The Collective as part of the extensive Old Oak and Park Royal regeneration zone in West London, one of the largest regeneration projects in Europe. Commissioned by the agency Create London (of which there is no space to detail their numerous and dubious roles in regeneration sites) the work was situated in the Old Oak and Park Royal Development Corporation that is the first of the Mayor of London's zones where the Greater London Authority becomes the planning authority for development overriding the local council's own planning committees. A long interview with People's Bureau on The Collective's website about their residency steers the conversation continually back to the benefits of co-living. It was very sad to see all this. We had long before asked the question who are the people that the People's Bureau refer to? We may add this to the question of what is collective about a property developer who call themselves The Collective. Whatever the radical history of the notion of the people and the collective, it is eviscerated here by two parties who both profit in different ways from those histories.

Two Quick Chapters in The Local Decomposition of the Elephant

In another turn of events locally, we would like to detail here some of the smaller complicities activated between art and business or even small business and large business. In 2018, we discovered that the new Hej coffee shop that had opened in Elephant Park directly on the site of the former Heygate had decided to display a large arty photo of the old estate as part of it minimal Scandinavian-style chic. Needless to say the photo was the usual aestheticized portrait of 1970's Brutalist architecture, all symmetries and no actual residents on show despite the image being photographed in the midst of residents struggles against decant and displacement. We took offence and began an online criticism of this poor taste that after some fun and games that included some other anonymous campaigners seriously messing with the cafe's Scandinavian hygge, Hej lost the battle and removed the photo from its cool walls. For us, this was a tiny battle in the bigger war but such battles stem from a fidelity to the long-term resistance plan to keep Heygate alive as a site and as a weapon against continued social cleansing of the area despite its demolition.

In August 2019, an unholy partnership of Wild in Art, a for-profit business who produce art trails that usually benefit charities, Andersen Books who publish David McKee's *Elmer the Elephant* books and Lendlease landed at Elephant Park. They placed ten fiber glass Elmers painted by schools and groups with the help of artists solely around the site of Elephant Park, one of them even outside Hej naturally. An accompanying Trail and Activities Map available for people to follow the trail did not disguise its primary function as slick marketing for Elephant Park. The first page talks about the development only as Elephant Park and Lendlease, despite being its developer, only appears as a logo on the back page. Sometimes social media is a good mode of attack as the publicity forces those complicit in this art-washing to have to answer for themselves. But this only works when the artwashers panic. Wild in Art and Andersen Press simply ignored our public accusations of their cynical and blunt instrumentalization of local schools and community groups. This not being some other more radical capital City, the Elmers were not set on fire and in the end some locals could only respond by leaving artfully crafted elephant dung behind the model elephants with accompanying signs that read 'Elmer Says I Don't Like Rich Fucks' and other choice local anti-gentrification slogans. Realistically such a micro-battle might point to an equation something like this: the farther away in time from the demolition of The Heygate you go, the less any artwasher will know about the ongoing community project that Heygate Lives! But it also points to the fact that each battle is a component of the whole, and these mini squibs and pissy protests should not be written off against other more material community struggles.

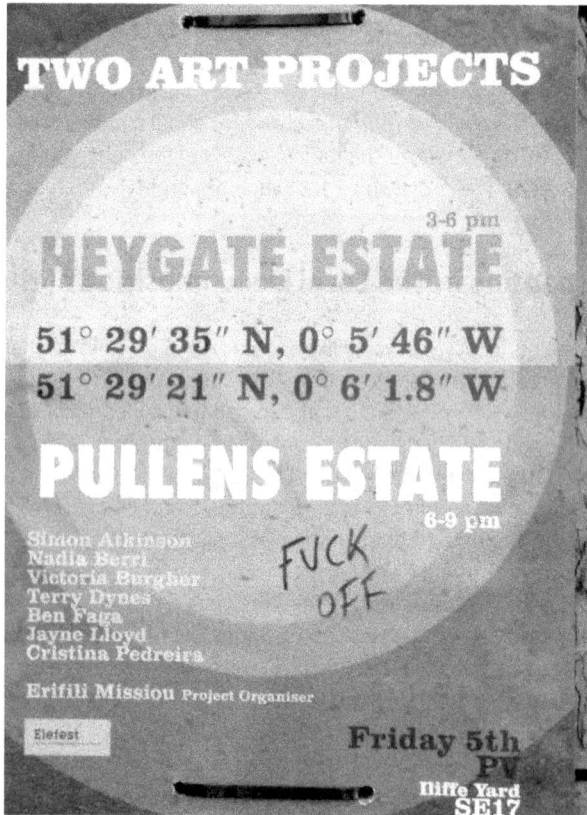

Fig. 3: Graffiti on local poster announcing another round of artistic happenings on Heygate Estate. Photo by Southwark Notes.

From these complicated vignettes of local art boosterism, bad faith and complicity and by way of heading to some tentative conclusions, we wanted to think about some of these things in the gentrification pie concerning art and artists, community and, if we have to, coffee shops. As we have seen time and time again, art and artists are commissioned, funded and lauded in development projects as delivering local community participation. Although we have outlined a few examples above, there are dozens more locally that have come and gone entirely inconsequentially for those local people who did not participate but who face further rounds of social cleansing. Reuben Powell, head honcho of the previously mentioned Hotel Elephant worked with Lendlease that saw local school pupils producing drawings of the demolition for the development's marketing hoardings. Artists Deborah Mason and Rebecca Manson Jones worked on the Aylesbury Estate in 2016 on a funded-project *Sticking Together SE17* as part of the Calouste Gulbenkian Inquiry into the Civic Role of Arts. They got local people from the Aylesbury, a much bigger estate that similar to Heygate is being demolished by Southwark's Labour council, to make collages about anything they wanted to assess whether poor people benefit from such engagements but not about who actually benefits materially from such funding. Our contention is that funding for these half-arsed bohemian forays into working class life would be better served by being given directly to tenants to produce their own culture as they see it. But this would never happen as this is not how the system maintains itself. Artists in London in particular remain spectacularly clueless on these issues despite twenty years of critical writing on art and regeneration. These art projects do little to actually enhance, act in solidarity with and fight alongside local communities. More likely these projects obscure and also delegitimize already existing local cultures and imaginations.

Investing in Your Flat White Life

Despite how much we don't like these artistic pseudo-encounters, for Southwark Notes, it is mostly foolish to equate freelance artists being willingly abused by business interests and new cool coffee shops as being some kind of real enemy. Hej Coffee, like others appearing in the neighborhood, are the product of gentrification and not the cause. More subtle and complicated are artists working for developers. Although we spend some time battling these small local contributions to the landscapes of gentrification, it would be stupid to center our *entire* political life on opposing them. Activist-only campaigns against *yuppies* in Docklands in the 80's and against arty hipsters currently in the East End or at The Elephant have little potential for actual community organizing and they seemingly don't have much longevity after the frothy anger has quickly dissolved! Activism is not organizing but organizing contains activism. In the same weeks that the mini-soap opera was playing out with us, Hej Coffee and others, we were also usefully spending our time being part of the local campaigns against the demolition of the Shopping Centre and of the Aylesbury Estate, because opposing these is opposing the much larger forces at play – that's national and local government-supported gentrification dressed up as *regeneration*.

But having said this, there *are* many communities in the world who are militantly against the arrival of gentrifying businesses and artists as they see them as part of the signal that it's okay for developers and speculators to colonize ever greater chunks of our cities for their bloody-minded profit-seeking schemes, while low-income residents get stomped over or cast aside.

In Boyle Heights in Los Angeles, a Latina neighborhood with a decades long struggle for public housing and against the social and ethnic cleansing of their area, parts of that community are targeting new business such as cafes or art galleries that are a sign of the violent gentrification of their area. Those struggles are enacted by a community that recognizes itself as such – as poor and as Latina – and who are fighting from the basis of that self-identification. As a community they are very clued up about how art is used to try and pacify opposition to and obscure the violent urban processes of speculation. They also have many artists working from within that community, not so much as artists but as people engaged in solidarity with the struggles because they live there and they feel it all too!

We would have to say that in London the debate and action around these questions has not been so clear and intentional. Of course, it's London and not L.A and each anti-gentrification struggle is different but it would be good to see some discussion in anti-social cleansing campaigns about how to think through and link different parts of the regeneration process. On the one hand there is global development and real estate, finance, investment and speculative development. On the other hand there is the 'uplifting' of areas with luxury flats full of wealthier residents and more expensive shops, different lifestyle values and displays, and endless artistic interventions or alleged social practice projects. But let's not mistake one for the other nor the scales of power involved. Freelance artists doing their engaged work or white coffee slurpers have little power over any of this, the same as the rest of us. We need to be clear also that many (but not all) people living in gentrifying areas are often likely to be as precarious in their housing situation (low wages and expensive rents), even in new developments like Elephant Park. Then of course, there are the rich folks buying into the area and so what role do they play as consumers of the regeneration / gentrification product and as those who replace those displaced by social cleansing? We have always questioned the pure praxis of fighting only the supply side or what we call the production side of process of social cleansing (developers, finance, local State, national State as the heavy structural players) and not fighting the demand side or what we term the consumption side (those buyers of new homes and cortados as consumers of newly-gentrified areas), not the least because we are still against the actual rich and won't be letting them off the hook from their tasteless and poor choices in consumption.

Getting out of That Dark Vertiginous Forest

In some ways for local people or groups it's easier to either ignore such artistic projects in regeneration or to be a bit pissy about them, or worse, to try and use them for your own propaganda and campaigning. For us it's clear that the enemies (and we think we need to see it that way) are always the local Labour council, Lendlease, Delancey or any non-profit or social enterprise who comes to the area to destroy it. You could see our and others guerilla war against artwashing as what it is – simply sticking to our faithfulness of the need to fight on all fronts but also the understanding that you don't need a massive mobilization against all of the time. Although locally housing campaigns have rarely involved themselves much in actually confronting these useless art projects or even publicly criticizing community members who take up roles of facilitating such arty adventures, they have been in support of how we have opened up more fronts to fight on.

But it is also true that in many ways the older local Elephant campaigns in calling for interim uses for the Heygate site had giving up on the idea of actually saving the estate and were only giving good future marketing ammo to Lendlease by organizing a community visioning event where part of the days desires for interim uses were taken up lock, stock and by barrel by the developer without it even having to pretend it was interested in local people's ideas. For us it was hard to stomach the sheer waste of organizing possibility when over 150 people came together to imagine and describe what they loved about their neighborhood and what they wanted to keep but this massive potential was frittered away on interim uses. But, in an often-mooted Southwark Notes conundrum, we were never clear when was the best time to advance a more radical criticism of this way of going about things. Would a moment of disunity that seeks a useful and practical critique go down badly or would it actually have opened up a much more complete understanding of what we are all up against at The Elephant? Looking back, we should have pushed local campaigns harder on the question of whether such support for interim uses was undermining the campaign work. We didn't and five years later we wonder why we didn't. Sometimes you've just gotta shake it and not worry so much about what you think you know the result will be. This is the rupture or the escalation that campaigns will always need to avoid becoming institutions or stuck in their ways. With this in mind we encourage all communities fighting social cleansing to go for broke and to find ways to fight on all fronts with the notion of *fighting to win*. Although we no longer have an answer to question 'What would it mean to win?', we recognize the long haul of struggle always. It's more likely to be usefully true that the struggle is always a win because if we didn't do that, we have already been defeated. We, and our friends across many local groups and campaigns, won't be stopping anytime soon.

You can find out more about these many struggles and our accompanying analysis on our lively blog:

Southwark Notes, https://southwarknotes.wordpress.com

FLIPPING THE SCRIPT ON ARTWASHING: FIGHTING GENTRIFICATION WITH TENANT POWER

SCHOOL OF ECHOES LOS ANGELES

We are talking to people about the galleries surrounding us and those who are thinking of coming in the future. The galleries need to understand how they impact housing and the neighborhood. They know very well that they will increase property values and rents. They will displace those of us who are poor and low-income. We don't see their type of business benefiting us. The galleries can go [back] to Hollywood or other places where they are not going to cause damage. I don't understand why they want to bring damage to our neighborhood and put us at risk.[1]

— Boyle Heights, Alliance Against Artwashing and Displacement

It is of critical importance to understand the gentrification process — and the art world's crucial role within it — if we are to avoid aligning ourselves with the forces behind this destruction. Definitions of gentrification — most generally issuing from the gentrifying classes-describe moments in the process, not the process itself. [...] For gentrification cannot be defined unless we first isolate the economic forces that are destroying, neighborhood by neighborhood, city by city, the traditional laboring classes.[2]

— Rosalyn Detsche and Cara Gendel Ryan

Much recent literature has attempted to define gentrification apart from its class dynamics. City officials, urban planners, developers and all their ancillary agents employ a rich vocabulary around gentrification, praising its so-called benefits and improvements, recasting gentrification as redevelopment or revitalization to mask what is inherently class violence. While gentrification was once synonymous *with* that violence, the term has come to signify something benign.[3]

From the perspective of communities targeted by dis-investment and reinvestment, however, there can be no accumulation without dispossession, no gains for real estate capital without the social cleansing of the poor and working class. For this reason, our collective, School of Echoes, chose to adopt a definition of gentrification that speaks from the perspective of those victimized by gentrification rather than its beneficiaries. Gentrification is the displacement and replacement of the poor for profit.

---

1   Boyle Heights Alliance Against Artwashing and Displacement, 'The Women of Pico Aliso: 20 Years of Housing Activism', http://alianzacontraartwashing.org/en/coalition-statements/the-women-of-pico-aliso-20-years-of-housing-activism/.

2   Rosalyn Deutsche and Cara Gendel Ryan, 'The Fine Art of Gentrification', October 31 (Winter, 1984), pp. 94.

3   We are borrowing Tom Slater's analysis of the changes in how gentrification signifies in urbanist discourses. See Tom Slater, 'The Eviction of Critical Perspectives from Gentrification Research', *International Journal of Urban and Regional Research* 30.4 (December, 2006), pp. 737-757.

The decades-long re-signification of gentrification has attempted to recalibrate social cleansing in the public imaginary as a social good. These efforts include gentrification's imbrication with transit-oriented development, tourism development, the development of public parks, *affordable housing* development, sports-oriented development, and what we will call arts-oriented development, or artwashing. In each instance, community groups that oppose speculative real estate and its gentrifying effects face accusations that they oppose positive neighborhood change. Groups that resist the construction of a bio-tech corridor are accused of opposing hospitals. Communities that oppose large-scale luxury housing and retail in a former working-class neighborhood are accused of opposing small business owners. And in the context of art and culture, communities that oppose the designation of an arts district as a prelude to luxury development are accused of hating art and artists.

Since the 1997 debut of the Guggenheim Art Museum in Bilbao, Spain, urban planners have prioritized the use of artists and arts institutions in the economic redevelopment of cities and neighborhoods disfigured by deindustrialization. Here art, artists, and art institutions are used both materially and ideologically to foster and protect gentrification.

Celebrating the so-called Bilbao effect, urban planners use the resources of private capital in concert with government incentives to instrumentalize contemporary art in focusing investment capital and flipping neighborhood demographics. Their plans range from capital-intensive art institutions, elaborate multi-site biennials and arts festivals, and the city planning designation of arts districts to more modest efforts like public arts commissions and art walks. Similarly, development boosters following Richard Florida have centered the *creative class* as the protagonist for neighborhood transformation at the expense of poor and working people. Here, the presence of artists and creative professionals is seen as effective in shifting the image and material reality of a neighborhood—from one marked by purposeful disinvestment, into one that has been up-cycled or cleaned up, and made attractive to middle class renters and buyers.

As creative class discourses and initiatives have aged with time, their impacts have come into sharper relief. Rather than discovery, we see erasure. Rather than revitalization, we see the death of community. Rather than creating supportive spaces for artists, we see luxury consumerism. Communities around the world have learned that arts-driven revitalization efforts depend on the displacement of long-term residents and the homogenizing of local culture for a culture of capital. Meanwhile, the brand of a creative class continues to sell grander scales of speculation long after the artists, designers, and other professionals have themselves been displaced and replaced.

The process of arts-oriented gentrification follows such familiar stages that planners and developers now view it as a formula to be replicated. What some argue is a natural or inevitable process, or the result of individual consumer choices, is, by now, highly centralized, controlled and scripted. Landlords, developers, private businesses, and city planners present arts-oriented gentrification as a win-win for the city and for business while ignoring the loss of home, employment, and way of life for poor and working class members of the community. In this article, we critically focus on the structural role of the arts as leverage within the processes of gentrification and explore how to build resistance against gentrification scripts.

Three Scripts of Gentrification

Since 2012, School of Echoes Los Angeles has conducted militant field research on the agents of gentrification and forms of resistance to it. Contrary to purely academic approaches to the problem of gentrification, the collective has consistently tested its analysis in political action. Those actions began with tenants rights workshops in the Echo Park neighborhood. The scale of the intervention shifted dramatically in the Summer of 2015 with the founding of the L.A. Tenants Union, an autonomous tenants movement for resisting social cleansing and building tenant power.[4]

Over the last five years, the L.A. Tenants Union has taken on dozens of battles. The union has organized building-based tenants associations and neighborhood tenant union chapters, or *locals*. One of the early struggles led by the Union de Vecinos Eastside Local of L.A. Tenants Union was a fight to stop the encroachment of gentrifying businesses in the Boyle Heights neighborhood surrounding the Pico Aliso public housing projects. A primary target of that campaign was a growing arts district.

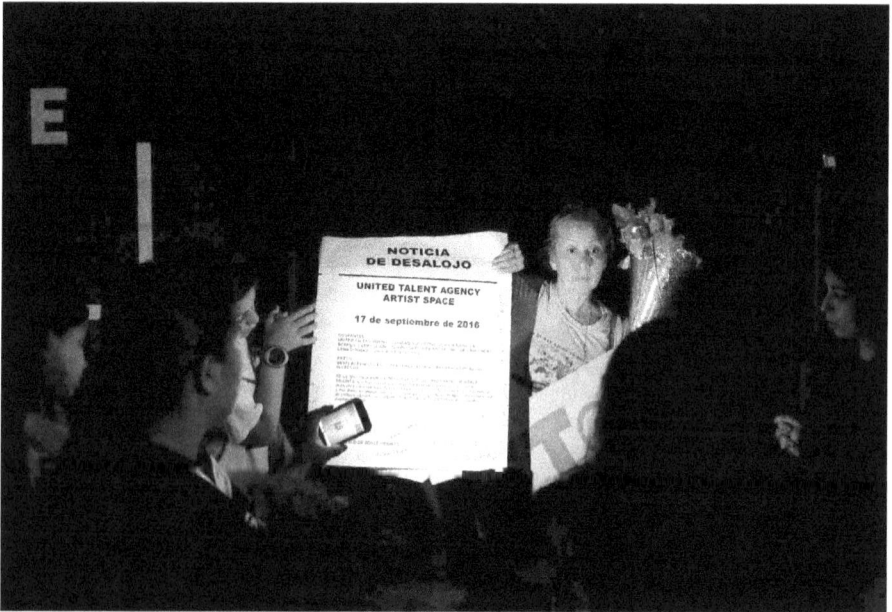

Fig. 1: United Talent Agency Artist Space eviction notice signed and posted by Boyle Heights resident Teresa Alfaro (right) and other members of the community, 17 September 2016. (Photo: Courtesy of BHAAAD)

---

4   The State of California Civil Code defines a tenant as persons who 'hire dwelling units.' Rather than limit the tenant to the system of rent, a political definition invokes the larger property relations under capitalism. Thus, in the context of L.A. Tenants Union and the larger tenant power movement, the tenant is defined as anyone who does not control their housing; including students in dormitories, the incarcerated, seniors in care facilities, single-room occupancy renters, lodgers. It also includes the homeless, or unhoused tenants. See Tracy Jeanne Rosenthal, '101 Notes on the LA Tenants Union', *Commune* 4 (Fall, 2019): https://communemag.com/101-notes-on-the-la-tenants-union/.

Since several School of Echoes members are also members of the Union de Vecinos Eastside Local, our collective had an insider's view of the Boyle Heights conflict. Much has been written about that struggle. It continues to raise crucial questions, not the least of which for artists. For School of Echoes, the battle against gentrifying businesses in Boyle Heights greatly advanced our understanding of how gentrification works and the ways that communities can defend themselves against it. The aim of analysis must always be to resist social cleansing and to defend the empowerment of poor and working-class communities in all their complexity, creativity, contradiction, and agency.

Fundamental to an analysis of resistance to gentrification is the conviction that, rather than a natural or inevitable process, gentrification is purposefully produced. Rather than a moral problem to be solved by changing individual behavior, gentrification is a political problem which must be resisted through collective action. Furthermore, that resistance must respond to the needs of poor communities and never undermine their autonomy, that is, it must occur within and alongside the long-term processes of organizing.

With the tenant power movement as our classroom, we have come to understand that real estate speculation depends upon what we call the three scripts of gentrification. Those scripts function in a broad range of gentrification processes, whether the focus is on transit-oriented development, or development that instrumentalizes tourism, entertainment, new technological industries, historic preservation, etc. The scripts used by real estate capital remain remarkably consistent. First, there is the image of the agent of gentrification as a pioneer or protagonist in the grand narrative of urban change. Second, gentrification depends upon representations of revitalization that counter or erase the consequences of social cleansing. And, third, since it is presented as a public good, gentrification requires public investment in the form of planning support, tax incentives, zoning changes, and publicly-financed infrastructure.

In blunt terms, the three scripts of gentrification entail a protagonist, propaganda, and public policy. Like a Gordian knot, all three of these scripts require each other in order for real estate capital to remake neighborhoods and cleanse communities without sparking riots and mass disruption. There is no public policy without the image of the heroic entrepreneur and urban pioneer. Without the discourse of revitalization, the gentrifying protagonist is exposed as a violent opportunist. In the following sections of this essay, we will review the three scripts of gentrification and apply those scripts to our experiences in resisting arts-oriented development or, to use the more common expression, art-washing.

The Script of the Urban Pioneer

When I came here, no one was here.

The urban pioneer script presents gentrifying entrepreneurs as a spatial vanguard, leading the way from historic dis-investment to development. Importantly, it includes a subjective investment that speculators and entrepreneurs have in the effects of historic disinvestment. It's the love of the empty warehouse as a *blank slate*.

The artist as an urban pioneer is by now a famous narrative in urban, suburban, and even rural contexts. In *Desert America*, Rubén Martínez described how land-art interventions of the 1960s echoed myths of colonial expansion, framing the desert as an empty place to which the settler (artist) arrives. More recently, Artist Andrea Zittel has described coming to Joshua Tree as arriving in a void. 'When I came here, no one was here.'[5]

In the desert and in the city, the word *pioneer* is often used romantically by artists and art institutions without acknowledging its colonial history. Artists promote a fantasy of settling new territories and discovering the frontier. Zittel calls it, 'the last edge'.[6] The myth of *uninhabited* space echoes the legacy of colonization, stitched into the American project from its inception. The erasure of indigenous history, culture, and indigenous peoples themselves is itself erased, the place presented as a *tabula rasa*.

The fantasy of the artist as pioneer follows the narrative of a personal quest. A move into a de-industrialized zone or the resource-starved streets lining public housing reflects back to the artist, not just a cheap place to live but an imagined space free from constraints on creative and lifestyle experimentation. In a social, cultural, and normative *vacuum*, the artist can invent new ways of living together. An expression of pioneering can occur even before any material action takes place. An artist enters a large, empty warehouse and feels a rush of elation, experienced as a sincere and honest feeling, imagining the ways that this property—a blank canvas—might be used. Rather than understanding the ways they participate in historical or strategic forms of disinvestment, the artist approaches the space as ahistorical, uninhabited, or *found*.

Here, artists betray their own subjective investment in and identification with their role as pioneers. They collapse the material conditions of affordability with their own fantasy of discovery. The erasure of historical consciousness through colonization, resource deprivation, exploitation, and redlining, is a precondition of that fantasy. The effacement of histories of structural violence becomes the ground against which the artist experiences freedom and liberation.

Within cycles of development, resource starvation in poor communities becomes a necessity for capital to create the conditions for future speculation. Such spaces may be referred to in the popular lexicon as *forgotten*. But the cycles of destruction are never accidental. Redlining

---

5   Rubén Martínez, *Desert America: Boom and Bust in the New Old West*. New York: Picador, 2019, pp. 79.
6   The same enthusiastic rhetoric can be heard in an April 2017 conference hosted by Claremont Graduate University and Sotheby's Institute of Art, titled 'L.A. as Lab: Extra Territories'. The conference brought together arts professionals, planners, and real estate developers to examine the ways 'Art and culture in Los Angeles are on the move, pushing past the conventional, beyond institutional frames, and out of bounds.' The conference website, https://www.cgu.edu/event/la-lab-2017-extra-territories/, features the conference organizers' response to the call for a boycott issued by Boyle Heights Alliance Against Artwashing and Displacement (BHAAAD). Due to the boycott, all but one featured artist as well as the program curator tasked with recruiting artists to speak at the conference withdrew from the proceedings. During the opening session of the conference a group of BHAAAD activists disrupted the event.

intentionally prevented people of color from accruing wealth through homeownership. Freeways were purposefully driven through communities with low-tax bases. Government resources are allocated to beautification projects in wealthy neighborhoods instead of basic infrastructure in poor ones.

The artist has a doubly-articulated role in the shift from under- to over-development. On the one hand, the role is ideological. The artist naturalizes their dependence on structural violence in order to access new territories. On the other hand, the artist's role is material. The artist who *arrives* precipitates the transition from destructive disinvestment to a new *creative* investment.

What is often left out of the story about the arrival of artists and art spaces are the policies and interventions that make such narratives possible. For decades, across Los Angeles and in cities throughout the United States, poor communities have borne the cost of redlining that suppressed homeownership and manipulated rental prices. The historic lack of public infrastructure such as sidewalks and streetlights, the purposeful placement of freeways, rail, and industry through communities all reveal decades of planned disinvestment in low-income and communities of color. The arrival of fine art spaces in these neighborhoods signals a change in fortunes for investors rather than for residents. The area has become ripe for speculation.

Unapologetically echoing the language of colonization and ignoring a history of embedded artists and cultural producers, an *LA Weekly* story about the Frogtown neighborhood in northeast Los Angeles congratulates Damon Robinson as 'the first contemporary artist-explorer' in the neighborhood. In 2007, Robinson arrived at the mixed-ethnic working class community wedged between the Los Angeles River and the Golden State Freeway and 'opened Nomad Art Compound, a sprawling warehouse that includes a bookstore, print shop, music venue, swimming pool and bedrooms for artists to rent.'[7] According to *LA Weekly*, Nomad Art Compound 'quickly became the foundation stone of Frogtown's resurgence.'

Behind what the writer calls *resurgence* are a series of government interventions that shaped the neighborhood. In 1959, Frogtown was connected to the whole of Chavez Ravine, a thriving, working class Latinx community. In 1960, the city evicted residents from their homes, promising to build a large public housing complex, where they would soon return. The promise was never kept. Instead, the land was handed over to private developers to build Dodger Stadium.[8] Just a few years later, the city chose the neighborhood as the location for the 5 freeway, isolating Frogtown from the rest of the community. Put in this light, Robinson is not a *pioneer*, but rather a beneficiary of government policy that systematically disadvantaged the residents of Frogtown and cut it off from surrounding neighborhoods.

---

7   Isaac Simpson, 'L.A.'s Hottest New Neighborhood, Frogtown, Doesn't Want the Title', *LA Weekly* (20
     August 2014): http://www.laweekly.com/news/las-hottest-new-neighborhood-frogtown-doesnt-want-
     the-title-5016257.
8   A telling omission that *LA Weekly* skips the mention of the promise of public housing as if the stadium
     alone forced the community's removal.

More, placing Robinson as the protagonist of the neighborhood's *resurgence* continues to erase the work (and artwork) of long-term residents.

Frogtown's gentrification was not simply a spontaneous process inaugurated by one artist but planned and produced. Soon after the opening of Nomad Art Compound, the city rezoned much of the area, borrowing from the 'creative class' planning playbook. By reclassifying the neighborhood from *manufacturing* to *commercial manufacturing*, the city made mixed-use and retail development possible. At the same time, the city passed gang ordinances over the area to give police more power to perform random searches, more discretion to judge activity suspicious, and a swift path to detain and arrest neighborhood youth.

The story of Frogtown's gentrification is often told from the perspective of the *pioneering artist*. The pioneering artist serves to naturalize as *blank canvas* the policies of dispossession, displacement, and neglect that have constructed disinvested neighborhoods. Featured as the protagonist of the neighborhood, the artist's perspective takes shape as the dominant perspective about the neighborhood: its history is erased and replaced with a fantasy of discovery. Finally, the artist serves to naturalize the material transitions of the neighborhood, hiding the planning and policies beneath the shifts.

The Script of Revitalization

Development catalyzes economic vitality and revitalization.

The script of revitalization represents gentrification as beneficial to historically underserved communities. In the field of art, representations of revitalization constructs art as an import brought into historically disadvantaged communities—mostly, poor and working-class communities of color. This construction of art ignores and de-legitimizes homegrown artistic production. Embedded cultural producers, including community artists, street artists, conceptual artists, crafts people, sign painters, musicians, and more are rendered invisible. Instead, art is seen as arriving into a deprived community in the form of white cube spaces and contemporary art spectacles authored by qualified professionals.

The script of revitalization relies on the perceived progressive benefits of art. If art is an inherent social good, the arrival of artists and the services that artists demand are accepted as harmless and non-controversial. Thus, the discourse around revitalization effaces the everyday needs of working-class neighborhoods, conceals the material impacts of gentrification, and discourages dissent. When revitalization schemes do acknowledge artistic and cultural vernaculars, it is still through the lens of outside economic development. Elevating or valorizing local cultural practices becomes a useful step towards demonstrating the viability of community development through speculative capital. Local cultural resistance to capital attains value only in the past tense when the targets are dead.

In an effort to expose the contradictions within arts-oriented development and its pretense to social good, activists have often employed the term artwashing. The term artwashing builds on analogous critical terms such as pinkwashing.[9] Pinkwashing refers to PR efforts to silence critique within an appearance of support for progressive interests. Activists have accused the Israeli government, for example, of reducing LGBT struggles to a public-relations tool. Here, pinkwashing is 'a deliberate strategy to conceal the continuing violations of Palestinians' human rights behind an image of modernity signified by Israeli gay life'.[10] Similarly, critics have referred to PR stunts with regards to environmental impacts of industry and development as greenwashing.[11]

London-based journalist Feargus O'Sullivan coined the term artwashing in 2014. 'When a commercial project is subjected to artwashing,' he writes, 'the work and presence of artists and creative workers is used to add a cursory sheen to a place's transformation.'[12] Artwashing operates at the level of public relations, advertising campaigns, planning, and cultural initiatives that promote development through art and culture. Artwashing is the discursive framework that upholds the positive impacts of the arts using language like *revitalization* or *community improvement*, while disguising the negative impacts of gentrification, particularly displacement.

At the citywide level, one example of artwashing in Los Angeles is Current LA, an arts biennial whose inaugural iteration centered around the Los Angeles River. The summer 2016 biennial was put on by the Los Angeles Department of Cultural Affairs but funded by Bloomberg Philanthropies, of technology billionaire, former New York mayor and presidential candidate, Mike Bloomberg. The $1 million in funding came from a Bloomberg program 'aimed at supporting temporary public art projects that celebrate creativity, enhance urban identity, encourage public-private partnerships, and drive economic development.'[13]

---

9   Though it describes a different operation of obscuring social reality, pinkwashing borrows from
     whitewashing, a term coined in the film industry. Whitewashing names the history of characters of
     color being played by white actors, as in the black face of *Birth of a Nation* or yellow face of *Breakfast
     at Tiffany's*. Whitewashing also refers to white actors taking roles on screen of historical or fictional
     people of color; e.g. Ben Affleck portraying Latino CIA agent Tony Mendez in *Argo* or actor Matt Damon
     cast as the hero of Chinese epic *The Great Wall*. See Amanda Scherker, 'Whitewashing Was One Of
     Hollywood's Worst Habits. So Why Is It Still Happening?' *The Huffington Post* (10 July 2014): http://www.
     huffingtonpost.com/2014/07/10/hollywood-whitewashing_n_5515919.html. For criticisms regarding
     the whitewashing of Asian characters, see Lawrence Yee, 'Asian Actors in Comic Book Films Respond to
     "Doctor Strange" Whitewashing Controversy', *Variety.com* (4 November 2016): http://variety.com/2016/
     film/news/asian-actors-whitewashing-doctor-strange-comic-book-films-1201910076/.
10   Sarah Schulman, 'Israel and "Pinkwashing"', *The New York Times* (23 November 2011): A31. See also,
     Tyler Lopez, 'Why #Pinkwashing Insults Gays and Hurts Palestinians', *Slate.com* (17 June 2014): http://
     www.slate.com/blogs/outward/2014/06/17/pinkwashing_and_homonationalism_discouraging_gay_
     travel_to_israel_hurts.html. For recent critiques of pinkwashing that promote regressive immigration
     reforms see Prerna Lal, 'How Pinkwashing Masks the Retrograde Effects of Immigration Reform', *The
     Huffington Post* (15 April 2013 and updated 2 February 2016): http://www.huffingtonpost.com/prerna-
     lal/pinkwashing-immigration-reform_b_3070788.html.
11   Joshua Karliner, 'A Brief History of Greenwash', *CorpWatch* (22 March 22, 2001): http://www.corpwatch.
     org/article.php?id=243.
12   Feargus O'Sullivan, 'The Pernicious Realities of "Artwashing"', *The Atlantic.com* (24 June 2014): https://
     www.citylab.com/equity/2014/06/the-pernicious-realities-of-artwashing/373289/.
13   Citation from the Bloomberg Philanthropies website announcing its financial support for Current LA,
     https://www.bloomberg.org/blog/public-art-challenge-the-winners-are/.

Throughout his tenure as Mayor, Eric Garcetti has led the charge for using the redevelopment of the Los Angeles River as a catalyst for gentrifying some of the poorest neighborhoods in the city. At the press conference launching the initiative he expressed enthusiasm for how Current LA would bolster his plans to *revitalize* the river corridor. The biennial took place at the same time that, according to *The Nation*'s Richard Kreitner 'more than half of riverfront properties have changed hands in the last three years [2013-16], sale prices have more than doubled, and rents have increased dramatically.'[14] The art biennial has the effect of washing over the changing demographics of Frogtown, a historically mixed-ethnic working-class community surrounded by freeways and the river basin. In Summer 2016, Garcetti also announced the hiring of celebrity architect Frank Gehry to spearhead the river *revitalization* effort. The same Frank Gehry who designed the Guggenheim Museum Bilbao.[15]

The Script of Public Investment

> *The Arts District is one of LA's most in-demand addresses, mere blocks from hip restaurants and gastropubs.*[16]

City-sanctioned arts districts are promoted as creating communities of artists, rather than opportunities for luxury consumption and speculation. Such schemes reveal, however, the role of local, state, and federal government in providing the financial, infrastructural, and policy conditions that make gentrification and arts-oriented development possible.

The role of city and state government is central to our analysis of arts-oriented development,[17] and involves the purposeful deployment of tax incentives, zoning variances, public investment in transit access and infrastructure, as well as the diversion of resources to police patrols and surveillance. Large-scale public investment for the designation of city-sanctioned arts districts has very different goals than supporting existing grass-roots artist communities. Rather, public investment in arts-based development acts as a means of driving speculation and economic growth. Cities use arts districts to draw in outside investment, attract tourism, and access funding streams from city, county, state, and federal pools. As a tool of urban planning, arts-districts help situate arts-oriented developments within working class neighborhoods or deindustrialized corridors.

14   Richard Kreitner, 'Will the Los Angeles River Become a Playground for the Rich?' *The Nation* (28 March-4 April 2016): http://www.thenation.com/article/will-the-los-angeles-river-become-a-playground-for-the-rich/.

15   As noted by Christopher Hawthorne in an *Los Angeles Times* Op-Ed, 'The quiet rollout suggests that River LA is less interested in giving a clear picture of what Gehry's plan eventually may include than in tamping down [...] worries that it may operate as a Trojan horse, a kind of high-design architectural cover, for rampant real-estate speculation in communities along the river.' Christopher Hawthorne, 'Frank Gehry's controversial L.A. River plan gets cautious, low-key rollout', *Los Angeles Times* (18 June 2016): http://www.latimes.com/entertainment/arts/la-et-cm-la-river-gehry-20160613-snap-story.html.

16   SunCal, '6AM Overview', http://suncal.com/our-communities/6am/.

17   For an analysis of how real estate capital has come to dominant urban planning, see Samuel Stein, *Capital City: Gentrification and the Real Estate State*. New York: Verso, 2019.

Arts-oriented development includes both for-profit and non-profit developments.

These can include new, high-profile museums, large-scale housing developments geared to arts professionals, boutique hotels, and shopping complexes tagged with terms like *art*, *creative*, and *design*. In many instances, arts-oriented development ventures require a prior or existing cultural history in order to boost its claim on the arts. Multi-cultural community development builds upon the prior existence of community-based galleries, and cultural ethnic institutions or practices. In other instances, planners and developers simply invent a cultural infrastructure as part of their development schemes.

Like transit-oriented development, arts-based development is promoted as community improvement, even as it eases the path for speculation and displacement. Transit-oriented development policies enable new mixed-use development close to public transportation exchanges, relying on government policy interventions: changing existing zoning policies, granting exemptions from planning regulations, and unlocking tax breaks. City officials push for transit-oriented development with the expressed intent of increasing density and expanding transit ridership. But, as UCLA and Berkeley research has shown, the impact of transit-oriented development tells a different story. Transit-oriented development acts as a catalyst for speculative development in historically disinvested neighborhoods, contributing to displacement and an overall decline in transit ridership.[18]

We use the term *arts-oriented development* to echo our analysis of transit-oriented development. Arts-oriented development depends upon similar public-private partnerships. It relies on the same government policy interventions: changing existing zoning policies, granting exemptions from planning regulations, and unlocking tax breaks. Both transit-oriented development and arts-oriented development rely on the rhetoric of community benefit while acting as a catalyst for high-end speculation. Rather than transit ridership or greener cities, the community benefit promoted by arts-oriented developments is a center for artistic community, or arts districts. Fundamentally, though, the most significant impact of

18   Transit-oriented development purports to link mass transit to high-density and mixed-use development projects. To incentivize investment, cities grant developers variances to zoning ordinances that would otherwise place restrictions on height, unit density, square footage, traffic, and parking requirements. The vast majority of transit-oriented housing consists of market-rate and luxury housing to maximize profitability. The upwardly mobile class of tenants contribute to the overall decline in mass transit ridership. In an absence of tenant protections, transit-development triggers waves of evictions in adjacent neighborhoods. Combined, these forces set off a chain reaction with immediate up-scaling of commercial real estate and/or the transfer of leases to corporate retail and corporate lease-holders. Transit development dramatically recomposes the class of residents in an area in addition to satisfying the needs of stakeholders in speculative development. See Elijah Chiland, 'Is Metro ridership down because low-income passengers are leaving LA?' *LA.Curbed.com* (22 May 2019): http://la.curbed.com/2019/5/22/18628524/metro-ridership-down-housing-gentrification-transit. Dick Platkin, 'Curious Facts about Transit Oriented Development (TOD) in Los Angeles Uncovered by CSUN Planning Students', *CityWatchLA.com* (10 January 2019): http://www.citywatchla.com/index.php/2016-01-01-13-17-00/los-angeles/16890-curious-facts-about-transit-oriented-development-tod-in-los-angeles-uncovered-by-csun-planning-students. Tracy Jeanne Rosenthal, 'Op-Ed: Transit-oriented development? More like transit rider displacement', *LATimes.com* (20 February 2018): http://www.latimes.com/opinion/op-ed/la-oe-rosenthal-transit-gentrification-metro-ridership-20180220-story.html.

arts districts is not cultural but economic: arts districts raise property values, spur speculation, and increase profits for flippers and property owners.

White noise from the freeway saturates the Pico Gardens housing projects. At the base of Boyle Heights, once referred to as the flats, Pico Gardens and Las Casitas are what remains of the largest concentration of public housing west of the Mississippi. Demolished under the Clinton Administration in 1996, 1,285 units of public housing in the Boyle Flats were reduced to just 296 units.[19]

In the last twenty years alone, Boyle Heights has seen 250 homes destroyed to build the Goldline Metro infrastructure and another 60 homes demolished by the expansion of the Hollenbeck Police Station in 2009. To the south, 1,175 rent controlled units on 93 acres were destroyed to build 5,000 market rate units. To the north, eight vacant lots owned by Los Angeles Metro will be flipped for Boyle Heights' first hotel and *affordable housing* development. Like most such developments, despite the tag, the housing that will be unaffordable to the vast majority of current residents. Finally, to the west, eleven art galleries have opened in the last four years, under the banner of a city-sanctioned arts district, inaugurated in 2012.

Boyle Heights has experienced a policy-sponsored wave of displacement in every decade of its history. In the 1930s, the U.S. government waged a *repatriation* campaign, deporting U.S. residents of Mexican descent to Mexico. According to Francisco Balderrama's *Decade of Betrayal: Mexican Repatriation in the 1930s*, one-third of Los Angeles's Mexican population was expelled between 1929 and 1944.[20]

Between 1940 and 1960, as much as 12% of the land available in Boyle Heights was turned into freeways. These federal-funded city initiatives led to the demolition of 2,000 homes and displaced 10,000 people. It is in the context of decades of resistance to privatization and displacement that local activists and organizers understand the designation of a Boyle Heights Arts District.

At the outset of arts-district designation, historic non-profit arts organization, Self-Help Graphics, held a series of meetings with community stake-holders to debate its merits and dangers.[21] But by 2013, art galleries and artists had already begun moving in. The galleries had the backing of a range of entrepreneurial interests, including blue-chip New York galleries and a few artist-run non-profits. Some had direct ties to real estate interests: the for-profit gallery

---

19   The Clinton administration's Housing Opportunities for People Everywhere (or HOPE VI) shifted federal resources for the maintenance of public housing into competitive block grants. Grants came with set of austerity measures including the stipulation that housing authorities had to demolish and rebuild housing projects to become less dense and accommodate mixed-income and private housing.

20   The repatriation campaign was fueled by fears that Mexicans and Mexican-Americans were taking scarce jobs and government assistance during the Great Depression. Of the more than 2 million people deported during this process, 60% were American citizens. See Francisco Balderrama, *Decade Of Betrayal: Mexican Repatriation In The 1930s*, Albuquerque: University of New Mexico Press, 2006.

21   Abe Flore, 'Promise, Pitfalls and the Boyle Heights Arts District', *KCET Artbound* (16 September 2012): http://www.kcet.org/shows/artbound/promise-pitfalls-and-the-boyle-heights-arts-district.

356 Mission (funded by New York's Gavin Brown) opened in 2013 with the encouragement of a developer; in 2016 the non-profit gallery PSSST was given free rent for twenty years by its building's owner, an anonymous investor. In all cases, the new galleries marked a profound shift from local arts organizations that emerged from and were designed to serve the interests of the long-term, low-income, and POC community.

The city's plan for the Boyle Heights arts district has extended much beyond securing artists and art spaces as tenants in the neighborhood. Here, the branding of art will help connect Boyle Heights to luxury retail and apartments Downtown. The new Sixth Street Bridge will cross the Los Angeles River and connect the Arts District in Little Tokyo to that in Boyle Heights. The 482 million dollar project, designed by Michael Maltzan, will incorporate parks, pedestrian walkways, and private art galleries.

A host of initiatives branded as arts-oriented are slated to transform the whole Los Angeles River basin. West of the river in the downtown arts district, a $2 billion development, under the name 6AM, will result in two 58-story towers, 1,305 market-rate apartments, 431 luxury condos, as well as an open-air mall with 23,000 square feet of 'art opportunity space'.[22] On the cement banks of the river itself, a development at 670 Mesquit Street is billed as artists housing with 800,000 square feet of office space, 250 residences, and two art boutique hotels. The 2110 Bay Street development will result in 50,000 square feet of commercial space, a 100,000 square-foot office building, and a tower of 110 live-work residential units for artists.[23] Bay Street promises to 'activat[e] a derelict area of the Arts District in Downtown Los Angeles.'[24]

Without prioritizing long term residents' ties to their homes and communities, the story of the Boyle Heights Arts District is bound to repeat the history of their implementation elsewhere. As property values rise in Boyle Heights, and as amenities keep pace with incoming wealthier residents rather than the needs of the long-term community, artists and art spaces will become the architects of mass displacement of long-term Boyle Heights residents and then of themselves. Rents will become too high; restaurants will no longer fit into their budgets; and the increasing luxury will be alienating even to them. In Boyle Heights, as elsewhere, to borrow from Martha Rosler, the arts district will serve not to support an artistic community, but pave the wave for the 'tranquility and predictably' which is needed to herald luxury developments.[25]

22   Andrew Khouri and Ben Poston, 'Towering development is proposed for L.A.'s Arts District: an "opportunity for density"', LATimes.com (24 September 2016): http://www.latimes.com/business/la-fi-arts-district-towers-20160924-snap-story.html.
23   A March 2017 profile in Architects Newspaper quotes Alan Pullman, senior principal architect for the 2110 Bay project as saying: '[We wanted to] design a project that felt like it was very connected to the existing character of the Arts District.' Antonio Pachco, 'Bombastic, brand-name architecture is transforming the L.A. Arts District', Architects Newspaper (9 March 2017): http://archpaper.com/2017/03/la-arts-district-brand-name-architecture/#gallery-0-slide-11.
24   Quoted from the 2110 Bay Street development promotional website, https://studio-111.com/portfolio/detail/2110-bay-street/.
25   Martha Rosler, 'Culture Class: Art, Creativity, Urbanism, Part II', E-flux 23 (March, 2011). http://www.e-flux.com/journal/23/67813/culture-class-art-creativity-urbanism-part-ii/.

Anti-Gentrification Strategies and Resistance

Discussions of gentrification have not always translated into politically meaningful action. Our analysis of its scripts and consequences seeks to change that. The scripts that we have laid out in the preceding sections—gentrifier as a pioneer, gentrification as revitalization, and the state-assisted policies of gentrification—help to lay the foundations for how we might think about intervening in these processes.

To begin with, it has become clear to us as organizers and popular educators that the only effective way of turning the tide on gentrification is to build a mass popular movement, such as a tenants' union. After five years of organizing within the L.A. Tenants Union we have learned the importance of centering the needs, aspirations, and values of the communities most directly affected by real estate speculation: low-income communities and poor communities of color. As our analysis has explored, speculative real estate interests frequently instrumentalize artists, art galleries, and arts institutions as elements of urban renewal strategies. Flipping the script on arts-oriented development requires interventions at every level of gentrification's protocols. That means organizing militant anti-gentrification movements, strategizing around counter-protagonists, counter-representations, and counter-forms of power.

Fighting arts-oriented development as one instance of the broader capitalist project of gentrification requires that movements take the scripts of gentrification and turn them inside out. Instead of flipping communities, movements flip the script: flipping the script on the artist as a pioneer to the protagonism of the poor; flipping the script on revitalization to self-determination and flipping the script on planned gentrification to building power. For the last section of this essay, we will tease out these transformations in the context of tactics developed by the tenant power movement in Los Angeles to resist arts-oriented development, gentrification, and developmentalism.

Fighting Against Galleries

The fight against gentrifying businesses, such as contemporary art galleries in Boyle Heights, has demonstrated the importance of selecting an appropriate symbolic target for any anti-displacement struggle. This target must serve as a visible reminder of the forces driving displacement. It must be public-facing enough to serve as the basis for a popular rallying cry. It must be malleable enough that, in resisting it, victory is conceivable and extends the horizon of future political possibilities. It must serve as an anchor around which a set of demands can coalesce, a new political identity can emerge, and a new commonsense can be built.

Oftentimes, it makes clear strategic sense to target precisely those actors and entities that are somewhat closer to us on the political spectrum. It is there that we might most successfully intervene in much larger processes and dynamics. Wealthy speculative real estate interests, including major city-backed developers, may well be the ultimate threat to community control in neighborhoods like Boyle Heights. Yet, those interests rely on entities like galleries and arts nonprofit organizations to clear the way for further speculative development by *breaking the seal* on neighborhoods that have historically been considered unsafe for speculative investment.

In the case of Boyle Heights, selecting the galleries as the initial targets of an anti-displacement fight proved to be strategic. We can synthesize much of what we learned in the context of this struggle into four central lessons. First, **expose the cracks in criticality**. There exists within contemporary art a longstanding interest in various forms of anti-capitalist and identitarian politics. These critical predilections can be used for political leverage. That is, if an art institution builds its brand on notions of criticality, radicality, queerness, intersectionality or even liberal notions of diversity, inclusion, or community legitimacy, then movements can more easily call their legitimacy into question when their stated positions contradict the consequences of their actions. When members of the community pull back the veil on the *critical* aspirations of an artwashing project, they undermine the cultural and social basis upon which such an institution depends. How long can a cultural institution that claims to be *inclusive* of *community interests*, or claims to advance radical or critical ideas, and so on, survive when its neighbors challenge that institution on precisely its own terms?

Second, **speak *through* the media.** The proximity of contemporary art galleries to circuits of media provide ample opportunities for press coverage. Proximity to media allows even a fledgling anti-gentrification movement to craft a powerful media narrative with a capacity to change prevailing political commonsense. The presence of freelance producers, curators, writers, and photographers within the artworld means that when there is direct action such as a picket-line or a blockade, artists guarantee press coverage. That coverage can appear in trade platforms like *Hyperallergic* and *Artforum* as well as in general outlets such as *LA Times*, *NPR*, *Newsweek*, or *The Guardian*.[26] While gaining widespread press coverage does not guarantee a movement's victory, it accelerates its conditions for success. Media attention broadcasts the moral crisis of social cleansing, transforming a local struggle into a national issue. It opens the possibility of crafting a new commonsense understanding of the issue, thereby creating the political conditions for victories or major concessions. Solidarity gives people with very little opportunity to distribute their analysis a means of speaking *through* the media to a larger audience.

Third, **boycotts force artists to take a position**. Galleries would not continue to exist without the ongoing consent of artists who show their work in them. Neither would galleries survive without the art handlers and preparators who provide material support for exhibitions, the arts writers who generate press, or the teams of support staff who keep institutions running. When an anti-gentrification movement persuades artists, in particular, to boycott galleries and pull their artwork from those galleries, then the demands of community members become difficult to ignore. The Boyle Heights struggle makes evident the beginning of a much broader realignment of the role of artists within the contemporary conjuncture. As we have come to understand, artists occupy an ambivalent role in the capitalist economy. As independent producers of goods, most artists are precarious workers. More often than not, artists secure their housing through the rental market—a fact that will only become more and more prevalent with the contraction of middle-class tenured teaching positions. However, because art functions as a luxury commodity in our society, some artists identify with the property-owning classes that collect art, those patrons are among the 1%, as well as large commissioning institutions like art foundations or

26   Saul Gonzalez, 'In this LA Neighborhood, Protest Art is a Verb', NPR KCRW (27 June 2017): http://www.npr.org/2017/06/27/534443389/in-this-la-neighborhood-protest-art-is-a-verb.

the State. Out of the demand for economic survival, artists have often associated themselves with the collector or philanthropic class, typically mediated through the middlemen of galleries and museums. Crucially, it is the same the collector class that profits from gains in real estate speculation. This situation provides dynamic organizing opportunities for an anti-gentrification movement that wishes to build cross-class power to stop the process of displacement through arts-oriented development. If a movement can successfully compel a realignment of artists' self-identification, then a powerful political bloc might emerge to hold a line against artwashing and displacement.

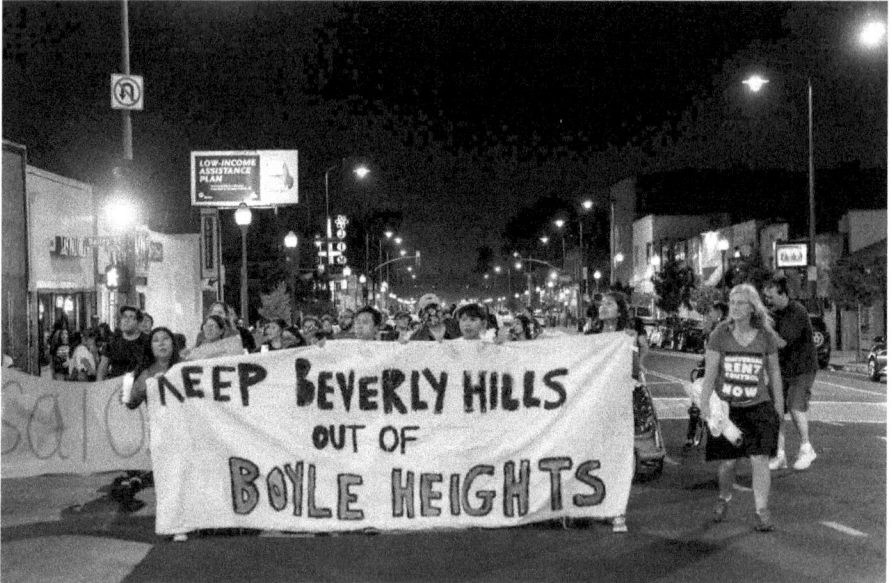

Fig. 2: Community members and supporters protest through the streets of Boyle Heights demanding that the galleries leave the neighborhood, 17 September 2016. (Photo: Courtesy of BHAAAD)

Fourth, **challenge the legitimacy of nonprofits**. Critics of the tenants fighting gentrification in Boyle Heights faulted the campaign for targeting nonprofit arts spaces over for-profit galleries. Much has been written about this tactic.[27] We will make two points on the matter. First, it's important to understand the structural relationships that nonprofit organizations have with communities and that for-profit institutions have with their nonprofit counterparts. In a situation where community members living in public housing had little power to stand up to large real estate interests and multi-billion-dollar planning processes, the nonprofit galleries occupied the weakest link in the arts-oriented development effort. Nonprofit organizations depend upon community consent in order to exist. If a significant bloc of community members withholds their support and reject the legitimacy of the organization, then that organization will struggle to secure funding. For-profit galleries, however, have no such dependence upon the community. They do, however, depend upon the existence of the nonprofit organizations in order to facilitate

---

27   See Ultra-red, 'Desarmando Desarrollismo: Listening to Anti-Gentrification in Boyle Heights', *Field* 14 (Fall, 2019): http://field-journal.com/issue-14/desarmando-desarrollismo.

community acceptance. Once the community undermines the legitimacy of the nonprofit organizations, the for-profit organizations are exposed to community resistance in all the forms that may take. It is no secret that nonprofit organizations depend upon the flows of capital, and real estate capital in particular.[28] In Boyle Heights, the well-respected and cherished community-based arts organization, Self Help Graphics, helped to promote the arts district planning through the advocacy of its board members. The board included specific individuals with direct material interests in the gentrification of the neighborhood: a well-known real estate investor in Boyle Heights and a lawyer representing real estate investors. Again, since the poor residents of public housing Boyle Heights had limited political power, they could threaten the reputation of the nonprofits by disrupting the idealist notions of art as a moral and social good apart from the mercenary interests of its financial backers who stand to gain financially from artwashing their real estate investment objectives in the neighborhood.

Fighting Against Displacement

If the artist as urban pioneer rests on centering the experience of incoming artists and arts institutions within historically disinvested neighborhoods, then fighting back requires that we center the experiences of community members who long called that neighborhood home. Who is the protagonist of gentrification? Real estate developers, state officials including planners and representatives, the mainstream media, as well as artists themselves highlight the protagonism of incoming residents. Who is the protagonist of the resistance to gentrification? We assert that it is the most vulnerable to gentrification's violence of displacement and harassment: namely, the poor, the houseless, predominantly marginalized people of color.

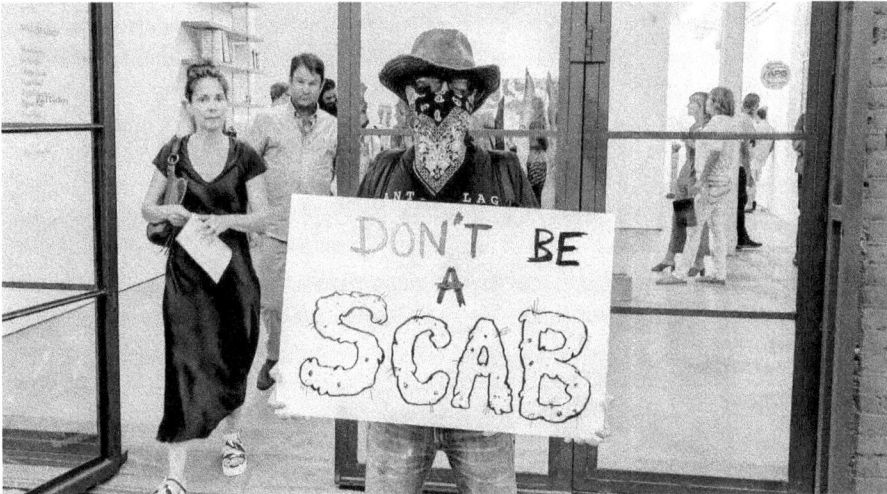

Fig. 3: Protestor on the picket line against Ibid Gallery, 15 July 2017. (Photo: Courtesy of BHAAAD)

---

28   For a revealing study of the links between philanthropy, artwashing, and right wing money, see Nizan Shaked, 'Looking the Other Way: Art Philanthropy, Lean Government, and Econo-Fascism in the United States', *Third Text* 33:3 (Fall, 2019), pp. 375-395.

At the same time, organizers must attend to the subtle class antagonisms even within marginalized communities. Capital's recruitment of an indigenous operative for outside investment is as much a part of the gentrification process as it is the history of colonialism. Local actors use the rhetoric of opposing outside interventions by activists and solidarity movements in order to then turn around and assume the authority granted to them by outside financial and political institutions. The slogan, 'gentify not gentrify'—which riffs on the Spanish word gente, meaning people—becomes a means of using identity-based representations to naturalize class relations within the community—class tensions that capital is all too happy to exploit (or, even invent). For tenant power movements, the protagonism of resistance must always be the subject position of the poor who stand to lose everything due to social cleansing, whether gentrification benefits outside interests or the landlords and entrepreneurs within the community itself.

Our analysis of gentrification must never lose sight of its real effects: displacement and replacement of the poor for profit. The central impact of arts-oriented development is displacement. A neighborhood that becomes a destination for one population becomes a site of dislocation for the poor. In the absence of legal protections that prioritize stable communities over the ability of landlords to turn a profit, tenants are displaced through rent increases, eviction loopholes, harassment, renovictions, Mickey Mouse repairs, or cash-for-keys scams. Unhoused tenants living in alleyways, parking lots, and sidewalks are displaced through dehumanizing sweeps.

As part of the tenant power movement, the L.A. Tenants Union organizes tenants. The movement prioritizes poor and working class people and the totality of their lives. We defend against arts-oriented development by fighting for the right to remain. We organize individuals, tenants in entire buildings, unhoused tenants in encampments, entire city blocks of tenants, and tenants united by shared property owners or management companies. The movement allows us to scale the struggle up to the city as a whole. The tools of our organizing work are many and comprise the daily efforts of L.A. Tenants Union: listening practices through popular education that expose the structural relationship between arts-oriented and other types of development with displacement. Those listening practices inform tenants' rights workshops, media training, organizers knowledge exchange, and door-knocking. Every strategy and every action becomes an opportunity to ask, what did we hear? Every fight for collective bargaining, rent strikes, pickets, or eviction blockades opens up spaces to reflect on what we heard. The process of critical reflection and the accumulation of knowledge transforms tenants in crisis into organic leaders. The movement is a classroom.

Fighting for Decommodification

If the script of revitalization relies on the perceived community benefits that come from art to wash away the material impacts of gentrification, then our movement must demand decommodification. If social cleansing betrays the fallacy of speculative 'development without displacement', then our movements must mobilize around self-determination and not further dependency upon real estate capital whose focus is on affordable housing scarcity, wasteful luxury development, and monopolization of ownership. In other words, the commodity housing system follows the same laws as globalized financialized monopoly capital. The anti-gentrification response is not community development that provides a Trojan Horse for speculative investment. We need to focus on what Michael Lebowitz has called, 'human development'.[29]

Long-term residents of poor and working-class communities have often demanded amenities which speak both to their needs and to long histories of capital flight. These tenants demand laundromats, schools, day care facilities, job programs, affordable grocery stores, clinics that provide affordable health, mental health, and harm reduction services. Community members demand fair banking services rather than exploitative check-cashing businesses. They agitate for general relief offices and spaces for community organizing and action. Arts-oriented development leaves these demands unmet. Instead of amenities that offer shared benefits with incoming residents, arts-oriented development creates unequal social geography in which long-time poor and working-class residents still cannot access affordable groceries and other daily services, while surrounded by cafes, boutique shops, and arts venues. The socially useful amenities that do exist in the neighborhood lose their commercial leases for gentrifying businesses. The corner market becomes an arts-supply store. The storefront church becomes a high-end photo-developing business. The auto garage becomes a craft brewery. The neighborhood fix-it shop makes way for cold-brew coffee.[30]

These contradictions reveal the falsity of arts-oriented development's claim to revitalize and beautify *blighted* neighborhoods. If for years community members have voiced the need for local businesses that provide basic services for everyday life, the struggle against gentrification holds within it the seeds of a vision of community where the needs of residents take priority over speculation. That vision includes collective networks of support that meet community needs, community ownership of local amenities, and community control over land-use decisions based on social use and not real estate exchange value. By sharpening the contradiction between gentrification's promise and its results, the movement demands decommodification against revitalization.

Fighting for True Public Safety

City officials tout arts-oriented development as a means of bringing public safety to crime-ridden neighborhoods. Artists and gallerists describe dangerous neighborhoods as having turned a corner. Streets that once had a reputation for vandalism, gang-on-gang violence, breaking and entering and other forms of harm are now perceived to be business- and pedestrian-friendly.

---

29   See Michael A. Lebowitz, *The Socialist Alternative: Real Human Development*. New York: Monthly Review Press, 2010.

30   For reporting that aided Los Angeles anti-gentrification activists' early understanding of the *science* of gentrification, see The Wealth and Poverty Desk, 'York & Fig At the Intersection of Change', *Marketplace* (1 December 2014): http://features.marketplace.org/yorkandfig/. The multi-segment public radio exposé details the role of real estate *retenanters* in changing the commercial landscape of the Northeast Los Angeles neighborhood of Highland Park. The article shows how real estate professionals use established templates regarding the types of businesses that invite higher-stakes real estate capital into working-class and immigrant neighborhoods. While the writers of the article ignore the role of city officials and finance, the article counters the fallacy that the arrival of galleries and other gentrifying businesses occurs organically or that it benefits poor and working-class residents. The aim is social cleansing.

However, the atmosphere of safety experienced by art patrons is not shared by long-term working-class residents. Often, community safety has a longer history of struggle and self-organization that precedes the arrival of artists and art spaces.

The struggle for public safety in poor communities, especially poor communities of color, has to be understood within the context of histories of police harassment, police brutality, and so-called quality of life policing. Historically, poor communities have learned that relying upon the police to settle disputes or to promote public safety results instead in an escalation of violence and further victimization. Racist policing practices around *broken window* policing, stop-and-frisk, three-strikes, and gang injunctions all serve to surveil and control poor communities of color through the cycles of capital disinvestment and re-investment. The experience of state violence often forces poor communities of color to organize their own networks to manage public safety autonomously, without the police. Autonomous organizing can minimize gang violence, control the flow of guns, transform alleyways, and inhabit public space at night. Crediting gentrifying businesses and higher income residents as responsible for neighborhood safety erases the years of community organizing in which the poor created public safety *against* criminalization and stigmatization.

But a formerly *blighted* neighborhood is never safe enough, not when tenants forced onto the streets roll their shopping carts down the sidewalk or the remnants of underground economies linger. Thus, at the same time arts-oriented development capitalizes on the labor of working-class communities to create a level of safety in the streets, gentrifying businesses continue to demand an increased police presence. Middle-class patrons and businesses call for increased surveillance and patrols, even when they encounter long-time community members, housed and unhoused. As marginalized communities have long known, increased police presence is experienced by long-term residents as harassment, incarceration, deportation, and even murder.

Resisting arts-oriented development requires that the tenant power movement understand community defense as defense *from* police and ICE agents and for community safety as produced through community organizing and mutual dependence.

Fighting for Belonging

Studying the impact of urban renewal on historic Black communities in cities like Pittsburgh, social psychologist Mindy Fullilove has described the community's experience of root shock. Families lose access to housing, jobs, schools, religious communities, and health services. Networks of mutual dependence are shredded. And individuals and families find themselves alienated from histories that sustain identity and wellness. For Fullilove, root shock becomes a fundamental condition of collective trauma. Gentrification, like urban renewal before it, triggers root shock through the shattering of community and the destruction of social practices developed by communities over time to organize social, economic, cultural, and psychological support in the face of poverty, institutional racism, and violence.[31]

---

31   Mindy Fullilove, *Root Shock: How Tearing up City Neighborhoods Hurts America, and What we can do About it*, New York: Ballantine Books, 2004.

Fig 4: Long-time residents of Pico Aliso public housing and leaders in the anti-gentrification fight in Boyle Heights, (left to right) Ana Hernández, Dolores Betancourt, Manuela Lomeli, and Delmira González. Seated in front the 356 S. Mission Road gallery boycotted by the community, 28 April 2018. (Photo: Courtesy of BHAAAD)

One essential role of the L.A. Tenants Union as a tenant power movement is to create and defend community, precisely where community is a target of gentrification's violence. Building solidarity between every person suffering from an eviction, or displacement, or too-high rent, or insecure tenancy, or landlord harassment, or slum conditions, or the inability to find an affordable home is not only a means to a more just world for tenants. It is also an end in and of itself. Shared meals, shared activities, regular meetings, are vital to our organizing because they intervene in the shocking alienation of gentrification, unifying community through mutual defense and support.

Fighting to Build Solidarity

In order to assume the scale as a citywide resistance to gentrification, a movement requires the involvement and investment of a wide array of actors from different class backgrounds, identities, and degrees of privilege. The L.A. Tenants Union has cultivated a culture of solidarity by framing housing struggles—including those of artists—as struggles of tenants and not of a particular professional group. Insisting that the subject of the struggle is the tenant, L.A. Tenants Union pursues an expanded vision of self-governance where tenants act upon a different vision for the city.

To resist arts-oriented development, our movement must invite artists and art professionals to form alliances with working-class residents on the basis of their shared identity as tenants and workers. Artists have a material need to recognize their own risk of displacement as gentrification processes accelerate. At the same time, artists have a political need to follow the leadership of poor and working-class tenants who have the greatest sense of precarity and the deepest connection to militant resistance while advancing the leadership of those whom society constructs are responsible for social reproduction, women. What would happen if artists took their identification as precarious tenants more seriously than their identification as artists? What if the politics of artists arose out of solidarity with and accompanying the struggles of the poor?

Solidarity means listening to those affected by the most violent aspects of contemporary life. People who live at the margins of society exist in a unique position to understand the threats that all tenants currently face and how best to solve those problems. Solutions that continue to replace and displace the poor for profit are not the answer to gentrification. If a tactic of resistance is rejected for being too *confrontational* or *impossible*, then whose material interests are being protected? When those of us who enjoy a level of comfort outside of the communities under threat of social cleansing and then publicly describe the tactics of those communities as 'terrorist' or 'extreme', then on whose behalf are we working? Such questions gain urgency since 2013 when the Department of Homeland Security issued a threat alert labeling anti-gentrification activists as 'anarchist extremists' posing a threat to local and federal law enforcement.[32]

Fighting to Take Over

The complicity of the state within gentrification's violence demands that we build an effective *outside* strategy. Such a strategy advances noncooperation, direct action, and autonomous organizing. For decades in Los Angeles, nonprofit corporations and liberal groups have pursued an *inside* strategy of negotiating with housing officials and policy makers behind closed doors. Community benefit agreements are negotiated in private between the executive directors of the community-based organization and developers. In order to ensure the goodwill or sympathies of public officials and developers, in order to safeguard private meetings and sanctioned participation processes, established non-governmental organizations have refrained from conflict and shielded officials and real estate interests from the anger of the grass roots. As the years passed, this compact between official community representatives, public officials, and capital has facilitated the continual negotiation of defeat. Nonprofits organizations capture the fear and anger of working-class and poor communities and direct that energy for the passage of community benefit agreements, density bonuses, and affordable housing. These same policies either directly contribute to or provide an alibi for the social cleansing of whole communities.

Systems of governance that serve the accumulation of private capital are not intended to preserve the best interests of the poor. They are designed to mute protest and marginalize militancy. For poor people of color, undocumented people, LGBTQ people, and others targeted by systems of structural oppression, the disproportionate ways in which the state metes out justice in our society is glaringly obvious. In the context of Arts-oriented

---

32   For a copy of the actual 23 July 2013 Department of Homeland Security memo see, http://publicintelligence.net/dhs-anarchist-gentrification-arson/.

development specifically, we understand that public officials, urban planners, and public policy, has a crucial role to play in flipping neighborhood demographics for the benefits of private accumulation of capital. Driving economic growth and raising property values has long been the central mandate for city officials, rather than the well-being of poor people. Without real veto power, officially sanctioned participation processes such as neighborhood councils and community planning hearings become shams. Where no never means *no*, community members expend their energy in consultation processes where the very existence of consultation satisfies the bare minimum of democratic requirements, regardless of the quality or outcome of that participation. A sign-in sheet is all the proof a developer needs to show that the community had its legally required hearing.

From the struggle for Civil Rights, the long march of farmworkers for dignity, the mobilization of queers and allies against government inaction in the face of the AIDS pandemic, to Black Lives Matter, the many successes of civil disobedience remind us that people in authority are not the only ones with the power to affect the course of our lives. Movements coalesce the testing of analysis with the convictions that change is made from the bottom up. Movements perform their unique character by embodying a strategy around tactics that dramatically refuse to cooperate or visibly withdraw consent.

Through protest, civil disobedience, boycotts, pickets, occupations, eviction blockades, and rent strikes, tenants win major victories against better-resourced real estate entities. Even without recourse to state intervention, movements of poor people can create the conditions that lead to concrete legislative gains through collective disruptive power. In the words of civil resistance scholar Erica Chenoweth, the goal of popular movements succeeds 'not by melting the heart of the opponent but by constraining their options.'[33]

Fighting to Reimagine the World

Embedded within the militant tactics of the anti-gentrification movement is the power to say 'no'. And 'no' means no, wherein communities claim the power to withdraw consent or to refuse cooperation with unjust laws, policies, or economic forces. At the same time, anti-gentrification movements built through tenant power affirm a collective subject. The tenant and the community of tenants have the potential to articulate a vision for the future of the city beyond the necropolis of commodified housing.

The process of building a tenant power movement that includes cross-class and cross-race solidarities is itself an instance of a radically new vision of the city, just as the creation of an organized community that can collectively determine its needs and then develop a strategy to meet them is not just a campaign. It is the formation of systems of interdependence, networks of solidarity, and longer-term processes of remaking the world. The fight for a more just and equitable society can have an enriching and transformative quality, as it can prefigure

33   Erica Chenoweth, 'It may only take 3.5% of the population to topple a dictator with civil resistance', *TheGuardian.com* (1 February 2017): http://www.theguardian.com/commentisfree/2017/feb/01/worried-american-democracy-study-activist-techniques.

the world to come. Organizing is not only a form of community defense; it is the process of collectively organizing a different world beyond what the given conditions say is possible.

For anti-gentrification militants oriented towards the collective making of a just world, it is not enough to reject one's complicity in gentrification. Stopping gentrification runs hand in hand with community defense. It involves defying neoliberal fatalism about the inevitability of class violence. It involves securing change that prioritizes human development over capitalist accumulation through displacement. Anti-gentrification in the practice of tenant organizing makes a more just world out of the material contradictions that presently exist. Defending and organizing community against the violence of social cleansing and its beneficiaries becomes urgent creative labor.

References

2110 Bay Street development promotional website, https://studio-111.com/portfolio/detail/2110-bay-street/.

Balderrama, Francisco. *Decade of Betrayal: Mexican Repatriation in the 1930s*, Albuquerque: University of New Mexico Press, 2006.

Bloomberg Philanthropies website, https://www.bloomberg.org/blog/public-art-challenge-the-winners-are/.

Boyle Heights Alliance Against Artwashing and Displacement. 'The Women of Pico Aliso: 20 Years of Housing Activism', http://alianzacontraartwashing.org/en/coalition-statements/the-women-of-pico-aliso-20-years-of-housing-activism/.

Chenoweth, Erica. 'It may only take 3.5% of the population to topple a dictator with civil resistance', *TheGuardian.com* (1 February 2017): http://www.theguardian.com/commentisfree/2017/feb/01/worried-american-democracy-study-activist-techniques.

Chiland, Elijah. 'Is Metro ridership down because low-income passengers are leaving LA?' *LA.Curbed.com* (22 May 2019): http://la.curbed.com/2019/5/22/18628524/metro-ridership-down-housing-gentrification-transit.

Claremont Graduate University and Sotheby's Institute of Art, 'L.A. as Lab: Extra Territories', conference website, https://www.cgu.edu/event/la-lab-2017-extra-territories/.

Department of Homeland Security memo (23 July 2013), http://publicintelligence.net/dhs-anarchist-gentrification-arson/.

Deutsche, Rosalyn and Rya, Cara Gendel. 'The Fine Art of Gentrification', *October* 31 (Winter, 1984): 91-111.

Flore, Abe. 'Promise, Pitfalls and the Boyle Heights Arts District', *KCET Artbound* (16 September 2012): http://www.kcet.org/shows/artbound/promise-pitfalls-and-the-boyle-heights-arts-district.

Fullilove, Mindy. *Root Shock: How Tearing Up City Neighborhoods Hurts America, And What We Can Do About It*, New York: Ballantine Books, 2004.

Gonzalez, Saul. 'In This LA Neighborhood, Protest Art Is A Verb', *NPR KCRW* (27 June 2017): http://www.npr.org/2017/06/27/534443389/in-this-la-neighborhood-protest-art-is-a-verb.

Hawthorne, Christopher. 'Frank Gehry's controversial L.A. River plan gets cautious, low-key rollout', *Los Angeles Times* (18 June 2016): http://www.latimes.com/entertainment/arts/la-et-cm-la-river-gehry-20160613-snap-story.html.

Karliner, Joshua. 'A Brief History of Greenwash', *CorpWatch* (22 March 22, 2001): http://www.
corpwatch.org/article.php?id=243.

Khouri, Andrew and Poston, Ben. 'Towering development is proposed for L.A.'s Arts District: an
"opportunity for density"', *LATimes.com* (24 September 2016): http://www.latimes.com/business/la-fi-
arts-district-towers-20160924-snap-story.html.

Lal, Prerna. 'How Pinkwashing Masks the Retrograde Effects of Immigration Reform', *The Huffington
Post* (15 April 2013 and updated 2 February 2016): http://www.huffingtonpost.com/prerna-lal/
pinkwashing-immigration-reform_b_3070788.html.

Lebowitz, Michael A. *The Socialist Alternative: Real Human Development*. New York: Monthly Review
Press, 2010.

Lopez, Tyler. 'Why #Pinkwashing Insults Gays and Hurts Palestinians', Slate.com (17 June 2014):
http://www.slate.com/blogs/outward/2014/06/17/pinkwashing_and_homonationalism_discouraging_
gay_travel_to_israel_hurts.html.

Kreitner, Richard. 'Will the Los Angeles River Become a Playground for the Rich?' *The Nation*
(28 March-4 April 2016): http://www.thenation.com/article/will-the-los-angeles-river-become-a-
playground-for-the-rich/.

Martínez, Rubén. *Desert America: Boom and Bust in the New Old West*. New York: Picador, 2019: 79.

O'Sullivan, Feargus. 'The Pernicious Realities of "Artwashing"', *The Atlantic.com* (24 June 2014):
https://www.citylab.com/equity/2014/06/the-pernicious-realities-of-artwashing/373289/.

Pachco, Antonio. 'Bombastic, brand-name architecture is transforming the L.A. Arts District',
Architects Newspaper (9 March 2017): http://archpaper.com/2017/03/la-arts-district-brand-name-
architecture/#gallery-0-slide-11.

Platkin, Dick. 'Curious Facts about Transit Oriented Development (TOD) in Los Angeles Uncovered by
CSUN Planning Students', *CityWatchLA.com* (10 January 2019): http://www.citywatchla.com/index.
php/2016-01-01-13-17-00/los-angeles/16890-curious-facts-about-transit-oriented-development-tod-
in-los-angeles-uncovered-by-csun-planning-students.

Rosenthal, Tracy Jeanne. '101 Notes on the LA Tenants Union', *Commune* 4 (Fall, 2019): https://
communemag.com/101-notes-on-the-la-tenants-union/.

Rosenthal, Tracy Jeanne. 'Op-Ed: Transit-oriented development? More like transit rider displacement',
LATimes.com (20 February 2018): http://www.latimes.com/opinion/op-ed/la-oe-rosenthal-transit-
gentrification-metro-ridership-20180220-story.html.

Rosler, Martha. 'Culture Class: Art, Creativity, Urbanism, Part II', *E-flux* 23 (March, 2011). http://
www.e-flux.com/journal/23/67813/culture-class-art-creativity-urbanism-part-ii/.

Scherker, Amanda. 'Whitewashing Was One Of Hollywood's Worst Habits. So Why Is It Still Happening?'
The Huffington Post (10 July 2014): http://www.huffingtonpost.com/2014/07/10/hollywood-
whitewashing_n_5515919.html.

Schulman, Sarah. 'Israel and "Pinkwashing"', *The New York Times* (23 November 2011): A31.

Shaked, Nizan. 'Looking the Other Way: Art Philanthropy, Lean Government, and Econo-Fascism in
the United States', *Third Text* 33:3 (Fall, 2019): 375-395.

Simpson, Isaac. 'L.A.'s Hottest New Neighborhood, Frogtown, Doesn't Want the Title', *LA Weekly* (20
August 2014): http://www.laweekly.com/news/las-hottest-new-neighborhood-frogtown-doesnt-want-
the-title-5016257.

Slater, Tom. 'The Eviction of Critical Perspectives from Gentrification Research', *International Journal
of Urban and Regional Research* 30.4 (December, 2006): 737-757.

Stein, Samuel. *Capital City: Gentrification and the Real Estate State*. New York: Verso, 2019.

SunCal. '6AM Overview', http://suncal.com/our-communities/6am/.

Ultra-red, 'Desarmando Desarrollismo: Listening to Anti-Gentrification in Boyle Heights', *Field* 14 (Fall, 2019): http://field-journal.com/issue-14/desarmando-desarrollismo.

Wealth and Poverty Desk. 'York & Fig at the Intersection of Change', *Marketplace* (1 December 2014): http://features.marketplace.org/yorkandfig/.

Yee, Lawrence. 'Asian Actors in Comic Book Films Respond to "Doctor Strange" Whitewashing Controversy', *Variety.com* (4 November 2016): http://variety.com/2016/film/news/asian-actors-whitewashing-doctor-strange-comic-book-films-1201910076/.

HEGEMONY AND SOCIALLY-ENGAGED ART: THE CASE OF PROJECT ROW HOUSES - HOUSTON, TEXAS, USA

ALYSSA ERSPAMER

A Case for Critical Art?

Toni Morrison is not alone in claiming that 'all good art is political'.[1] But what does such a statement mean? *How* is it political, and *what* 'political' is being referred to? I believe that the Gramscian concept of hegemony offers us a valuable lens with which to answer these questions. Socially-engaged art, specifically the renowned artwork Project Row Houses, will be my guinea pig, or case study, for the application of this lens.

Socially-engaged art is a genre currently popular with art institutions (e.g. see New York-based organization A Blade of Grass and the Andy Warhol Foundation's 2018 grant recipients) that continues to elude exact definition. Along with relational aesthetics, it decisively broke with the art object and its organizing criteria around the end of the 20th century.[2] However, as Suzanne Lacy establishes in her seminal work, socially-engaged art (her *new genre public art*) differs from relational aesthetics in its explicit valorization of the political, considered at the core of its mission and identity.[3] Its similarities with institutional frameworks have led to its projects being ascribed an 'NGO-style'.[4] Intersecting both politics and art, socially-engaged art is a fertile ground for hegemonic analyses.

The concept of cultural hegemony comes from Marxist philosopher Antonio Gramsci. Gramsci's reflections on the successes of communist revolution in Russia compared to its failures in the West led him to acknowledge the critical role played by the superstructure within the political realm, something the Marxist structuralist theory he ascribed to tended to underestimate. With his hegemonic theory, Gramsci carved out a space in evolutionary Marxism for unexpected results, such as the lack of communist revolution in the West. In the West, would-be revolutionaries were not merely confronted with antagonism from the state apparatus, but also from their surrounding *civil society*, which was far stronger than in the East. Civil society, or the spaces of social and political life outside of the state, is the political terrain on which both hegemony and its resistance are organized.[5] Hegemony can

1   Kevin Nance, 'The Spirit and the Strength: A Profile of Toni Morrison', *Poets&Writers*, (November/
     December, 2008), https://www.pw.org/content/the_spirit_and_the_strength_a_profile_of_toni_
     morrison.
2   Nicolas Bourriaud, *Relational Aesthetics*, trans. Simon Pleasance and Fronza Woods, Dijon: Les Presses
     du reel, 2002; Suzanne Lacy, 'Introduction', in Suzanne Lacy (ed) *Mapping the Terrain: New Genre
     Public Art*, Seattle: Bay Press, 1995, pp. 19-47.
3   Lacy, *Mapping the Terrain*, pp. 26–29.
4   Tania Bruguera and Larne Abse Gogarty, 'Citizen Artist', *Art Monthly* 400 (October 2016), p. 5.
5   David Forgacs (ed.) *The Antonio Gramsci Reader: Selected Writings 1916-1935*, New York: New York

119

thus be understood as the dominant social, political, ideological, and economic system that both the state and its citizens (consciously and unconsciously) ascribe to and reproduce. It is what, in any given time and place, we consider to be *normal*. Hegemonic beliefs also serve to justify the underlying economic structure of society. Their defeat is not as simple as an armed revolt against an authoritarian state, but rather requires what Gramsci terms a *war of position* (from the trench warfare of World War I): a drawn-out, taxing struggle that challenges hegemonic articulations at every societal node, both symbolically and materially, slowly revealing the ideological and determined nature of our reality in order to unveil the existence of alternative ways of life.[6]

Hegemony is propagated by states as well as individuals: by their actions and unquestioned belief in its dominance. Gramsci was reacting to capitalism; since his time, capitalism has evolved and adapted to a world of globalized finance, weakened industry and union power, and widespread precarity. Neoliberalism, as we call capitalism's most recent mutation, has different, at times more pervasive, technologies of power and control, yet it remains vulnerable to hegemonic analyses, as both Chantal Mouffe and Nancy Fraser have convincingly demonstrated.[7]

Political philosopher Chantal Mouffe has contributed greatly to bringing hegemony into the contemporary art discourse. She sees society as a battleground of hegemonic articulations vying for control — whenever one dominates, it masquerades as the natural, rather than determined, state of affairs.[8] Art, she writes, has the potential to 'create little cracks' in the smooth facade of neoliberalism (our current hegemonic articulation) and show how 'things could have been otherwise',[9] even as hegemony appears to deny us any choices outside itself. She terms art that actively challenges the status quo 'critical',[10] and considers it a break from the traditional avant-garde that used to offer the radical critique within art. The avant-garde can too easily be co-opted now; in order to stay political, artists must be more *counter-hegemonic* than innovative. They must help construct 'new subjectivities'[11] as alternatives to the accepted subjectivities of our status quo.

Political art thus has the potential to be a counter-hegemonic project. I will focus on a specific artwork: the well-known, socially-engaged Project Row Houses (PRH), celebrating this year (2019) its 25-year anniversary. PRH exists in the space of dozens of renovated shotgun houses and several other properties in the Third Ward neighborhood of Houston, Texas, and includes rotating art installations, afterschool education programs, and a single young mothers' program, among other elements. A new website and public identity accompanied the recent anniversary, and PRH now defines itself as a 'community platform that enriches lives through

University Press, 2000, pp. 222–24.
6   Forgacs, *The Antonio Gramsci Reader*, 226.
7   Nancy Fraser, *The Old is Dying and the New Cannot Be Born*, London: Verso, 2019; Chantal Mouffe, 'Artistic Activism and Agonistic Spaces', *Art & Research* 1.2 (2007): 1-8.
8   Mouffe, 'Artistic Activism and Agonistic Spaces': 3-5.
9   Chantal Mouffe, 'Alfredo Jaar: The Artist as Organic Intellectual', *Revista Diseña*, 11 (July, 2017): 22, 29.
10   Mouffe, 'Alfredo Jaar: The Artist as Organic Intellectual': 23.
11   Mouffe, 'Alfredo Jaar': 8.

art with an emphasis on cultural identity and its impact on the urban landscape'.[12] Artist Rick Lowe is the figure most associated with the project, as he was largely responsible for initiating it in 1994, and has remained connected to it since. Lowe, an African American artist from rural Alabama, has worked and lived in Houston since the mid-1980s, and is now based in the Third Ward. Inspiration for PRH came in part from a tour he took of the neighborhood in 1992, when he saw rows of abandoned shotgun-style houses and perceived their beauty and local significance;[13] and in part from a high school student's provocative question asking why he didn't provide *practical* solutions through his art.[14] Lowe, along with a group of fellow artists, conceived of PRH in 1992 and started refurbishing several shotguns with the help of local volunteers and corporate sponsors. Through private and public funding, they pieced together PRH, including the exhibition space, a vegetable garden, and several units of affordable housing.[15] My rationale for focusing on PRH is its sustainability and reputation: 25 years is an impressively long time for a socially-engaged piece, and PRH has received wide-spread praise from art and media institutions alike – *ArtNews* claiming that it is 'changing the world'.[16] It has never been analyzed from a hegemonic perspective.

I believe that PRH presents several convincing and powerful counter-hegemonic elements, but that ultimately it fails to consolidate them into a counter-hegemonic totality. This is because PRH's output remains unconnected to material realities; is not presented as a critique to the current status quo; and does not link to broader struggles of resistance. I realize that in our Western cultural context, overwhelmingly controlled by capital and the equation of money with value, it is difficult for artworks to avoid neoliberalism. However, critical art must still aim to challenge the assumptions and effects of our current hegemony. PRH, instead, has been readily digested by the dominant belief system because its acts of resistance are not adequately contextualized or connected to a broader struggle. While PRH provides space for the imagining of a *better*, even *different*, world, this imagination ultimately remains devoid of any political or economic implications, and thus cannot present a true challenge to the ideological and material conditions preventing any such *better* world from emerging.

Confronting Hegemony Through Art...

If hegemony is perpetuated at every node in civil society,[17] then every node is also a potential source of resistance. This may be particularly true for architecture, as it often functions as

---

12   [About PRH], *Project Row Houses,* 2019, https://projectrowhouses.org/about/about-prh/.
13   Stephanie Smith, *The Art of Place and the Place of Art at Project Row Houses,* Master of Arts, Rice University, Houston, Texas, 1998, pp. 32–33.
14   April Garnet Sommer Rabanera, *Success Stories: An Exploration of Three Non-profits Working in Preservation, Affordable Housing, and Community Revitalisation,* Master of Heritage Conservation, USC School of Architecture, Los Angeles, California, 2013, p.105.
15   Sheryl Tucker, 'Reinnovating the African-American Shotgun House', *Places* 10.1 (1995): 64, 67-68.
16   Tom Finkelpearl, *What We Made: Conversations on Art and Social Cooperation,* Durham: Duke University Press, 2014, pp.144-147; Carolina Miranda, 'How the Art of Social Practice is Changing the World, One Row House at a Time', *Artnews,* 7 April 2007, http://www.artnews.com/2014/04/07/art-of-social-practice-is-changing-the-world-one-row-house-at-a-time/.
17   Ernesto Laclau and Chantal Mouffe, *Hegemony and Socialist Strategy: Towards a Radical Democratic Politics,* 2nd edition, London: Verso, 2014, pp.122–123.

the spatial articulation of hegemony: a rarely noticed, seemingly pre-existent form, which is actually created at a specific time by a specific ideology imposing, or at the very least suggesting, a specific way of being. As Phil Hubbard asserts, summarizing Bourdieu, 'architecture is a form of representation which naturalises certain meanings in the interest of certain groups'.[18] An idea confirmed by PRH itself, in its 49th artistic 'Round', or exhibit, which cites architect-activist Leslie Kanes Weisman: 'space, like language, is socially-constructed... The spatial arrangements of our buildings and communities reflect and reinforce the nature of gender, race, and class relations in society'.[19] Architectural space is thus an important field in which to disseminate counter-hegemonic activity and symbolism.

Fig. 1: Project Raw Houses Studios, September 2012. Photo by Hourick.

The shotgun house is the defining image of PRH, housing it physically and serving as its logo. A shotgun is a one-room wide and long wooden construction endemic to the southern United States but thought to have originated in West Africa and been brought over with slavery. Shotguns exist in repetitive rows with front porches and a common backyard space, encouraging social interaction among inhabitants. They have always been associated with African Americans, but after desegregation, as more affluent individuals moved away from their historic neighborhoods, they became linked with *impoverished* African Americans specifically.[20] Across their social lives, these houses have come to represent a specific

18   Phil Hubbard, 'Urban Design and City Regeneration: Social Representations of Entrepreneurial Landscapes', *Urban Studies*, 33.8 (1996): 1446.
19   [On View], *Project Row Houses*, 2019, http://projectrowhouses.org/on-view.
20   Edward Orlowski, 'House of Blues: The Shotgun and Scarcity Culture in the Mississippi Delta', in: Nnamdi Elleh (ed) *Reading the Architecture of the Underprivileged Classes: A Perspective on the Protests and Upheavals in our Cities*, London: Routledge, 2016, pp.79–80, 93–94, 96.

underprivileged ethnic group – their presence in a given space is therefore not neutral. They should be seen as symbols as well as habitations. By placing them at the center of its project, PRH is elevating the associated demographic to a similar position of centrality, symbolically uplifting what hegemonic reality has largely ignored. In this way, PRH is contributing to a counter-hegemonic war of position.

In fact, before Lowe saw the shotguns that would become PRH, they were in poor condition and slated for demolition.[21] This speaks to a crucial problem in the neighborhood and country at large: gentrification. The nearby Fourth Ward has experienced widespread displacement due to its central location,[22] with the Third Ward in the same incoming line of development. Gentrification is an insidious instrument for the prevailing hegemony, as it can eliminate alternative ways of life and their accrued symbolism (such as communal low-income African American communities), all while utilizing the language of positive regeneration. Its architectural style has been dominated for the last half-century by postmodern constructions often more *spectacular* than the buildings preceding them, placing, as Harvey claims, 'image over substance'. Harvey argues that an 'achievement' of such architecture is the veiling of socio-economic issues through a kind of playful distraction meant not to undermine the 'coherence' of the ruling paradigm (i.e. hegemony).[23] Shotguns are particularly noteworthy, then, in their aesthetic opposition to such architecture: their small stature, more traditional materials, and long history refuse assimilation into the postmodern architectural narrative. Resisting, and aesthetically inverting, ongoing gentrification and its sole concern with returns on investments is an example of a counter-hegemonic positionality in PRH.

The preservation of the shotgun houses in PRH represents, moreover, an act of *maintenance*. This in broader terms opposes the neoliberal insistence on constant innovation and growth, where Houston, for example, is seen as part of the 'expressway world', in which the old is bulldozed to make way for the new.[24] In art, the concept of maintenance is rooted in Mierle Ukeles' 1969 manifesto defending the work of maintainers (mothers, cleaners, etc.) in comparison to the more revered creative work of (often male) artists. Ukeles calls maintenance 'the life instinct' and the avant-garde 'the death instinct'[25] – we can detect echoes of Mouffe here, in the rejection of avant-gardism as a worthy ambition for art. Ukeles' concept can be expanded to the maintenance of a community – its values, its aesthetics – against incoming development: the maintenance of the Third Ward and its architecture, for example. In contrast to today's starchitect-led vanity projects, seeking to unveil a (decontextualized) architectural performance the world has never seen (think Elbphilharmonie, Hamburg), PRH's shotguns endorse that which already exists, but lies at the peripheries of the hegemonically-sanctioned.

---

21   Smith, *The Art of Place and the Place of Art at Project Row Houses*, pp. 33-36.
22   Rabanera, *Success Stories*, p. 120.
23   Cited in Hubbard, 'Urban Design and City Regeneration', pp. 1444–45.
24   Walter Hood and Carmen Taylor, 'Musing the Third Ward at Project Row Houses: From Cultural Practice to Community Installation', *Cite*, (Spring, 96): 30-33; Smith 1998: 39.
25   Mierle Ukeles, 'Maintenance Art Manifesto', in: Kristine Stiles and Peter Selz (eds) *Theories and Documents of Contemporary Art: A Sourcebook of Artists' Writings*, Berkeley: University of California Press, 1996 (1969), pp. 622–24.

I thus interpret the predominant position of shotguns within PRH as a counter-hegemonic artistic act, challenging both dominant architectural narratives as well as the neoliberal process of gentrification, utilizing the houses' accumulated history, aesthetics, and relation to their community to fashion a symbolic resistance within civil society.

Within the architectural realm, PRH has not only maintained, however, but also *developed* – developments which, I argue, can also be interpreted counter-hegemonically. In 2003, PRH founded the Row House Community Development Corporation (RHCDC), which is explicitly focused on housing issues through its provision of low-to-moderate-income affordable housing for a select number of local inhabitants.[26] The RHCDC combines this material attention with a desire for its housing to be 'creative', to preserve the 'character and architecture of the area', and to 'create community'.[27] Its houses are *updated* shotguns: wider and taller, they nonetheless maintain the aesthetics and communal features, such as porches, of the local architecture. The *new*, in this the case, is an endorsement of the old. John Vlach, a researcher on shotguns, claims, 'it is an integral part of the process of African American art to constantly reshape the old and familiar into something modern and unique to [...] reinforce the image of the community'.[28] The RHCDC thus exists within an artistic heritage of re-users, reacting to the moment's needs rather than vainly seeking the new (the danger of which is co-option, exemplified by the Bauhaus becoming Ikea).[29]

The work of African American, Chicago-based artist Theaster Gates, a prominent name in socially-engaged art, similarly looks to past and present, rather than future, for inspiration. In his Dorchester Projects and Rebuild Foundation, Gates' art is the community buildings he creates from renovated pre-existing structures and repurposed material.[30] Both Gates and Lowe utilize strategies of reuse and recontextualization that respond directly to the neighborhoods in which they work – for Gates, the Greater Grand Crossing of Chicago.[31] This brings me back to Mouffe and her appeal to artists: to respond to *specific* issues in *specific* places (specific hegemonic nodes), basing their work more around need than concept.[32]

An important political feature of the RHCDC's context-based architecture is its challenge to the dominant conception of what social housing is and looks like. Among affordable housing units, the RHCDC is unique. The most tangible difference is aesthetic as, in the United States, public imagination on social housing is still captured by the 'projects': tall, isolated blocks

26 Rabanera, *Success Stories*, pp. 115–119.
27 Row House CDC, https://www.rowhousecdc.org/.
28 Cited in Orlowski, 'House of Blues', p. 94.
29 Ben Davis, 'A Critique of Social Practice Art: What Does It Mean to be a Political Artist?', in: Johanna Burton, Shannon Jackson, and Dominic Willsdon (eds), *Public Servants: Art and the Crisis of the Common Good*, Cambridge, Massachusetts: MIT Press, 2016, pp.431.
30 Tim Adams, 'Chicago artist Theaster Gates: "I'm hoping Swiss bankers will bail out my flooded South Side bank in the name of art"', *The Guardian*, 3 May 2015, http://www.theguardian.com/artanddesign/2015/may/03/theaster-gates-artist-chicago-dorchester-projects.
31 Natalie Moore, 'How Theaster Gates Is Revitalizing Chicago's South Side, One Vacant Building at a Time', *Smithsonian*, December 2015, http://www.smithsonianmag.com/innovation/theaster-gates-ingenuity-awards-chicago-180957203/.
32 Mouffe, 'Alfredo Jaar: The Artist as Organic Intellectual': 25.

that started falling out of favor towards the end of the 20th century.[33] The contrast between them and RHCDC's human-sized, social housing is immediate: while the 'projects' were often ideologically and erroneously vilified, they do not imply the same degree of site-specific and respectful care that the RHCDC constructions do. Moreover, the latter has several practical features hard to find within traditional social housing. They are Energy Star certified and feature efficient air conditioning and insulation.[34] They are ultimately *not* cost-efficient[35] – indeed, their value seems to lie largely in their symbolic stance, functioning counter to the hegemonic norm. Importantly, this stance is in no way divorced from or superior to their material reality: rather, their material reality as functioning, well-designed social housing *is* a dissenting act, opposing the predominant material reality among social housing elsewhere.

Chilean architect Alejandro Aravena and his practice Elemental have also worked on reimagining social housing.[36] When the government was unable to provide funds to build individual, fully functioning houses for squatters in Northern Chile, and the squatters threatened to go on hunger strike rather than live in the proposed housing blocks, Elemental suggested a solution: building the squatters *half* of a fully functioning house each – complying with the minimum legal requirements and including all basic infrastructure – and then letting them finish the rest themselves.[37] Such a housing solution both respects the inhabitants and their political desires, and also recognizes their unique personalities and creative potential.

In both Elemental and the RHCDC's cases, the artists and architects seem conscious of their inability to structurally provide a whole community what it needs or *deserves* (i.e. affordable housing for all); they are instead concerned with producing living and material symbols of what *should* be done, what probably *could* be done if the ruling system had different priorities and placed inhabitants' well-being above profit. Through their existence, the RHCDC houses reveal that they *can* exist – this might seem a tautology, but in a hegemonic context, reality and possibility merge: the dominant hegemonic articulation must eliminate any alternative, for it seeks to be source and arbiter of all possibilities. It is precisely here that PRH and art in general have the opportunity to be political and create counter-hegemonic alternatives, suggesting the determined nature of *reality* and creating in those who experience such alternatives 'the desire for change'.[38] This desire awakens only when there is space for its growth, and it is not faced with continuous rejection from the dominant system. Artists could thus task themselves with nurturing the spaces and ideas that allow the desire for change to grow.

---

33 Jeff Crump, 'Deconcentration by Demolition: Public Housing, Poverty, and Urban Policy', *Society and Space* 20 (October, 2002): 582–84, 586.
34 Rabanera, *Success Stories*, p. 140.
35 Michael Kimmelman, 'In Houston, Art Is Where the Home Is', *The New York Times*, 17 December 2006, http://www.nytimes.com/2006/12/17/arts/design/17kimm.html.
36 Oliver Wainwright, 'Chilean architect Alejandro Aravena wins 2016 Pritzker prize', *The Guradian*, 13 January 2016, http://www.theguardian.com/artanddesign/2016/jan/13/chilean-architect-alejandro-aravena-wins-2016-pritzker-prize.
37 Ariana Zilliacus, 'Half a House Builds a Whole Community: Elemental's Controversial Social Housing', *Arch Daily*, 24 October 2016, http://www.archdaily.com/797779/half-a-house-builds-a-whole-community-elementals-controversial-social-housing.
38 Mouffe, 'Alfredo Jaar: The Artist as Organic Intellectual': 20.

In addition to its architectural offerings, I believe PRH has another important counter-hegemonic feature: its Young Mothers Residential Program (YMRP). This program provides subsidized housing, counselling, mentorship, and afterschool help to a group of young, low-income mothers in the neighborhood for one to two years.[39] The demographic of the overwhelmingly African American YMRP[40] is one of the most disenfranchised in the area and country. One in three single-mother families in the United States lives in poverty and a majority of them are African American; among women, African Americans have the highest percentage in poverty.[41] These statistics should come as no surprise in a country still living through the implications of its slave-owning past, with a dominant social and economic class consisting predominantly of white, Christian (overwhelmingly not single-parent), affluent men. PRH's chosen demographic is thus an inversion of the hegemonic elite.

The power of the YMRP is its enactment of alternative success stories, of a temporary space wherein its female members, who receive little confirmation from society at large, are respected and in control. As Guillermo Gómez-Peña of radical performance art troupe La Pocha Nostra claims, while art projects can create 'an imaginary space, we also know that it actually exists, even if only for the duration of a project'.[42] Herein lies the power of emancipatory experiments like the YMRP – their emancipation and enactment of alternatives are real, even if short-lived,[43] and perhaps sufficient to create new subjectivities: cracks in the smooth facade of hegemony. Hegemony's survival depends on its posing as invincible, as if pre-dating the very terrain of resistance, and thus even transitory alternatives like the YMRP may prove effective in implying hegemony's fragility in the minds of those who are crucial to maintaining it: members of civil society.

Another frame through which to understand the YMRP is artist Joseph Beuys' concept of social sculpture, an inspiration for Lowe. Beuys claimed that everyone is an artist and that simple social interactions can constitute art if approached creatively.[44] The YMRP aims to provide such a platform for creative and ameliorative practices, prioritizing interactive and experimental activities for participants to 'identify with and appreciate' their 'creative' selves.[45]

39 Rahanera, Success Stories, pp. 114–15.
40 Smith, The Art of Place and the Place of Art at Project Row Houses, p. 6.
41 Zenitha Prince, 'Census Bureau: Higher Percentage of Black Children Live with Single Mothers', Afro News, 31 December 2016, http://www.afro.com/census-bureau-higher-percentage-black-children-live-single-mothers/; Margaret Simms, Karina Fortuny, and Everett Henderson, 'Racial and Ethnic Disparities Among Low-Income Families', The Urban Institute, August 2009, https://www.urban.org/sites/default/files/publication/32976/411936-Racial-and-Ethnic-Disparities-Among-Low-Income-Families.PDF; Jasmine Tucker and Caitlin Lowell, 'National Snapshot: Poverty Among Women & Families, 2015', National Women's Law Centre, September 2016, http://nwlc.org/wp-content/uploads/2016/09/Poverty-Snapshot-Factsheet-2016.pdf.
42 Elena Marchevska, 'Gómez-Peña and Balitronica Gomez, from La Pocha Nostra troupe', in Elena Marchevska, the displaced & privilege: live art in the age of hostility, London: LADA, 2017, p. 57.
43 Brian Holmes, 'The Artistic Device, or, the Articulation of Collective Speech', Ephemera, 6.4 (2006): 426.
44 Cara Jordan, 'The Evolution of Social Sculpture in the United States: Joseph Beuys and the Work of Suzanne Lacy and Rick Lowe', Public Art Dialogue, 3.2 (2013): 146, 144, 150.
45 Young Mothers Residential Program, Project Row Houses, 2017, https://static1.squarespace.com/static/55832a9de4b0920b55b16891/t/588ba5c186e6c0ffe0e5d4c3/1508433636183/

The importance of interactions in a social sculpture lies in the moment they are carried out, during which individuals can come to see life as creative and malleable, and thereby become more receptive to change. Embodied alternatives to our expectations and experiences help suggest that what appear as inevitable patterns are actually products of certain dominant ideologies, and that alternatives to our reality *can* exist.

...Without the Political Foundations

Despite all these convincing elements, I ultimately do not consider PRH, in its totality, a true or effective counter-hegemonic artwork. And I doubt PRH itself, including Lowe, would either. Its goals do not appear either explicitly or implicitly counter-hegemonic. This does not diminish the importance of my analysis, which does not seek to reflect PRH's interests or goals, but rather to provide a tool through which to gauge the political within art. Though PRH's work may embody and/or represent positive potential, moments of exception and reversal, community pride and beauty, none of these elements seems intent on drawing deliberate connections with a wider, and determining, hegemonic system. I do not mean to suggest that all art *must* be critical, or that PRH's output is thus a waste, but rather only to illuminate PRH's role within a sanctioned and accommodating artistic discourse, one which prefers to place a creative and ethical spin on reality than to consider its critical foundations.

PRH is ultimately satisfied with the status quo, and therefore we may want to ask ourselves whether the features explored above might, instead of presenting counter-hegemonic potential, be rather artwashing mechanisms for the current hegemony. Artwashing is the practice of using the arts or creative activities to give a positive or ethical guise to the activities of neoliberalism (or other practices). Although neoliberal actors have helped the arts in many instances, artwashing implies that any resulting artistic production becomes co-opted into the larger machine that keeps neoliberalism alive and unquestioned. As Ben Davis claims, the 'robber barons' of capitalism can excuse themselves from their actions through their cultural sponsorship.[46] Mouffe also draws attention to late capitalism's reliance on the creative industries for *valorization*[47] through the appropriation of the symbols and narratives these industries produce.

Perhaps the simplest place from which to start is PRH's rather unsavory sponsors, including Chevron (a sponsor since PRH's inception) and, until recently, Bank of America.[48] In a neoliberal context where capital and meaning are inherently entangled, sponsorship is no neutral act. The logo and visibility of sponsors on the projects they fund function as a sort of symbolic appropriation of that content and its own symbolism. Chevron, for example, is one of the world's largest energy companies, especially for petroleum – its role in environmental destruction and the propagation of neoliberal hegemony is self-evident. PRH itself has

YMNRP+Application+2017.pdf.
46   Jordan, 'The Evolution of Social Sculpture in the United States': 432.
47   Mouffe, 'Artistic Activism and Agonistic Spaces': 1.
48   Ben Davis. 'A Critique of Social Practice Art: What Does It Mean to be a Political Artist?', in: Johanna
      Burton, Shannon Jackson, and Dominic Willsdon (eds), *Public Servants: Art and the Crisis of the
      Common Good*, Cambridge, Massachusetts: MIT Press, 2016, p.432.

acknowledged these implications in one of its artist rounds, Round 44: 'Shattering the Concrete: Artists, Activists and Instigators'. Curator Raquel De Anda spoke of the danger of energy extraction and the importance of 'mass popular social movements like the Climate Justice Movement. She stated, moreover, that art can play a role in 'challenging our current political paradigm and [...] inviting [communities] to participate in altering the conditions that shape our lives'.[49] Though these claims appear to support a hegemonic understanding of the society in their challenge to the current order and recognition of the need for mass movements, they are undermined when Chevron is in the background, funding its own *debasement*. De Anda also exhibited works from the artist collective *The Natural History Museum*, which led a successful campaign to remove David Koch from that museum's board of directors,[50] recognizing the power, and danger, of such sponsorship within neoliberal society. Thus, PRH endorses an anti-hegemonic logic whilst simultaneously transforming it into rhetoric through its own choices and sponsors.

This argument also extends to the recent sponsorship of Bank of America, to which Davis, writing on PRH, gives special attention: he highlights that it is one of the organizations that most profited from the foreclosures of millions of homes in the 2008 crisis.[51] PRH, meanwhile, attempts to sell itself as a champion for lower-income homeowners, investing money (and profiting in reputation) in the RHCDC to ensure that parts of the neighborhood remain out of private hands. This mission may seem like a joke, or at least a mere propaganda scheme if we consider Bank of America's role. Bank of America has a much wider impact on homeownership than PRH; PRH may thus be surrendering any possible symbolic advantage by selling out its integrity. In terms of scale, the current hegemony may even be gaining more from PRH's valorizing and washing up of Bank of America's reputation than suffering from its symbolic critique of the current housing situation.

With low public funds available and civil society increasingly marketized, it may seem impossible to avoid corporate sponsorship completely or to create creative production untainted by capital. Yet, sponsorship aside, PRH's output still falls short of the counter-hegemonic. While PRH may encourage certain participants or viewers to believe that a better world is possible, this *better* is devoid of political or economic connotations: it is internal to the artworld and to the Third Ward. The same could be said of Gates or Aravena's work – and indeed, the sustainability of these artists' approaches is questionable. Aravena's half-built houses now sell for five times their original price,[52] while continuing processes of urban gentrification in Chicago and Houston put both the Third Ward and Greater Grand Crossing specifically at risk.[53] Thus *the system*, or the various processes we term neoliberalism, could

49   Raquel De Anda, 'Round 44: Shattering the Concrete: Artists, Activists and Instigators', *Project Row Houses*, (Spring 2016), https://static1.squarespace.com/static/55832a9de4b0920b55b16891/t/56f43 d45b6aa60e521becfb5/1458847054898/Round+44+booklet.pdf.
50   De Anda, 'Round 44: Shattering the Concrete: Artists, Activists and Instigators': 1–2.
51   Davis. 'A Critique of Social Practice Art', p. 432.
52   Wainwright, 'Chilean architect Alejandro Aravena wins 2016 Pritzker prize'.
53   Joy Sewing, 'Gentrification of the historic Third Ward', *Houston Chronicle*, 2 August 2018, https://www. houstonchronicle.com/business/real-estate/looped-in/article/Listen-Gentrification-of-the-historic-Third-Ward-13124478.php; Corilyn Shropshire, 'Chicago gentrification fears rise as East Garfield Par,

end up erasing what these projects have achieved. The community members of PRH and the Dorchester Projects are fortunate to be, for the moment, within these spaces. Yet these communities' general disenfranchisement and lack of guarantees – the reasons, indeed, why these artworks seem so necessary – remain unquestioned by the art, which begins to seem like superficial, symptomatic care.

Following Mouffe, PRH's approach might be understood as a political *third way*, referring to centrist political parties that actively avoid veering either to the left or right. These parties, according to Mouffe, are politically unfeasible, for they produce a facade of consensus which ignores the hegemonic nature of society and denies publics the necessary agonistic and active choice between competing alternatives.[54] PRH may similarly be seen as attempting to appease all sides, producing certain counter-hegemonic symbols whilst receiving hegemonic stamps of approval. This strategy cannot disarticulate the current hegemony because it is supporting its existence. Moreover, it creates the impression of a cohesive project (and reality), which in actuality is dictated by antagonism: the needs of PRH's *public* (or participants), for example, versus the actions of its corporate sponsors. This antagonism is never publicly acknowledged.

As Mouffe claims, no art project seeking to be counter-hegemonic can afford to act in isolation, for 'it would be a serious mistake to believe that artistic activism could, on its own, bring about the end of neoliberal hegemony'.[55] That end depends on the recognition that there are multiple nodes of resistance in civil society and that they must grow conscious of each other, creating a maze of trenches, or 'chains of equivalence', that only together offer a convincing and powerful new counter-hegemonic articulation. Yet PRH does not engage with external activist groups or counter-hegemonic artists; it does not attempt to participate in any such chain.

It is important now to return to the source of my discussion around cultural hegemony: Gramsci. In his own writing, hegemony is an explicitly political phenomenon, product of a Marxist line of thought: an explicatory tool for modern capitalist and democratic societies' complexities and resistance to revolution. Gramsci, however, never denied society's economic base structure. As a Marxist, he believed that the source of the superstructure – of ideology and hegemony – lay within the economic base structure, limiting his argument to defending an analysis of the former also on its own terms.[56] Artworks like PRH are distant from this political lineage, interacting little, if at all, with the underlying economy that produces the components of PRH's social sculpture. Reconsidering the YMRP, it seems unfortunate that it does not explicitly and politically challenge the material foundations of society, as the difficult circumstances of its female participants are not bad luck but rather *products* of a dominant economic paradigm. Without considering these foundations, it is difficult for an artwork to

Austin, South Lawndale housing prices increase', *Chicago Tribune*, 20 December 2018, https://www.chicagotribune.com/business/ct-biz-chicago-housing-costs-gentrification-worries-20181219-story.html.

54   Mouffe, 'Alfredo Jaar: The Artist as Organic Intellectual': 22.
55   Mouffe, Chantal. 'Artistic Activism and Agonistic Spaces': 5.
56   Laclau and Mouffe, *Hegemony and Socialist Strategy*, pp. 58–59.

be truly critical of its effects. In a sense, the YMRP's *alternative success stories* support the myths of the prevailing system, as they imply that its effects are surmountable through self-application and creativity.

A Case for Critical Art

So, what is the political in art? In our neoliberal society, I believe art gains its political dimension when it addresses our prevailing hegemony and aspires to be *critical*, following Mouffe, by challenging the status quo. PRH, one of the most celebrated and long-lasting works of socially-engaged art, is a suitable starting point for this analysis in part because of its pre-eminence and in part because of the difficulty in assessing it on purely aesthetic grounds as a socially-engaged piece.

The basis of my analysis is Gramsci's theory on cultural hegemony, which posits that the dominant obstacle to Marxist evolution in today's capitalist democratic societies exists within their superstructure: specifically, their *hegemony*, or the reigning ideology that is produced and reproduced not only at the state level, but also at the level of civil, or private, society, and that serves as continuous justification for the conditions of the underlying economic structure.

PRH, despite its accomplishments within the artworld and the Third Ward, does not, ultimately, present a critical artwork geared at providing counter-hegemonic alternatives. Its potentially counter-hegemonic production provides an example of how socially-engaged works may begin approaching the challenge of affronting hegemonic neoliberalism. Through its staging of *emancipatory experiments* such as the YMRP, and its powerful and alternative use of architecture, PRH offers a potential dissent. However, its critique is never fully articulated: it is never tied in with other struggles or with hegemony's political and economic lineage. It avoids addressing the origins of some of its own concerns, like gentrification and the economic marginalization of certain demographics. Playing instead with the reversal of end effects, PRH may ultimately be furnishing neoliberalism with cultural cushioning and sanctioning – and thus assuring the reproduction of its hegemonic assumptions. To avoid such a fate, PRH would have to be more critical of its relationship with capital and eliminate sponsors such as Chevron; provoke discussion and creative production related to the more foundational conditions of society; actively link up with radical projects and groups working in similar areas; and continue offering spaces for the creation of new subjectivities, such as the YMRP.

I do not intend to discard PRH and its work, to label it a failure, or to ignore the beneficial effects it has on many of its participants. Instead, I seek to bring to light the political question of hegemony and how artworks, in an era of increasing encroachment of capital into the culture, may understand themselves in relation to it. Nor do I view neoliberalism as a monolithic, inflexible entity. It is rather precisely neoliberalism's flexibility and localized forms[57] – which nonetheless invariably prioritize the expansion of capital over

---

57 David Harvey, *Breve Storia del Neoliberismo,* trans. Pietro Meneghelli, Milan: Il Saggiatore, 2007, p.85.

human well-being – that require an active and attentive response from the cultural and artistic spheres. As Mouffe has argued, artists should not be so discouraged by the threat of neoliberal appropriation as to retreat outside the workings of society; this would accomplish little in terms of resistance.[58] Rather, they must stand firm in their belief in culture and art's potential to shock and reveal, to suggest and stage alternatives. They will then be able to contribute to the struggle for the articulation of our next, and better, hegemony.

References

[About PRH], *Project Row Houses*, 2019, https://projectrowhouses.org/about/about-prh/.

Adams, Tim. 'Chicago artist Theaster Gates: "I'm hoping Swiss bankers will bail out my flooded South Side bank in the name of art"', *The Guardian*, 3 May 2015, http://www.theguardian.com/artanddesign/2015/may/03/theaster-gates-artist-chicago-dorchester-projects.

Bourriaud, Nicolas. *Relational Aesthetics*, trans. Simon Pleasance and Fronza Woods, Dijon: Les Presses du reel, 2002.

Bruguera, Tania and Larne Abse Gogarty. 'Citizen Artist', *Art Monthly* 400 (October 2016): 1-5.

Crump, Jeff. 'Deconcentration by Demolition: Public Housing, Poverty, and Urban Policy', *Society and Space* 20 (October, 2002): 581–596.

Davis, Ben. 'A Critique of Social Practice Art: What Does It Mean to be a Political Artist?', in: Johanna Burton, Shannon Jackson, and Dominic Willsdon (eds), *Public Servants: Art and the Crisis of the Common Good*, Cambridge, Massachusetts: MIT Press, 2016, pp.423-435.

De Anda, Raquel. 'Round 44: Shattering the Concrete: Artists, Activists and Instigators', *Project Row Houses*, (Spring 2016), https://static1.squarespace.com/static/55832a9de4b0920b-55b16891/t/56f43d45b6aa60e521becfb5/1458847054898/Round+44+booklet.pdf.

Finkelpearl, Tom. *What We Made: Conversations on Art and Social Cooperation*, Durham: Duke University Press, 2014.

Forgacs, David (ed.). *The Antonio Gramsci Reader: Selected Writings 1916 1935*, New York: New York University Press, 2000.

Fraser, Nancy. *The Old is Dying and the New Cannot Be Born*, London: Verso, 2019.

Harvey, David. *Breve Storia del Neoliberismo,* trans. Pietro Meneghelli, Milan: Il Saggiatore, 2007.

Holmes, Brian. 'The Artistic Device, or, the Articulation of Collective Speech', *Ephemera*, 6.4 (2006): 411-432.

Hood, Walter and Carmen Taylor, 'Musing the Third Ward at Project Row Houses: From Cultural Practice to Community Installation', *Cite*, (Spring, 96): 26–33

Hubbard, Phil. 'Urban Design and City Regeneration: Social Representations of Entrepreneurial Landscapes', *Urban Studies*, 33.8 (1996): 1441–1461.

Jordan, Cara. 'The Evolution of Social Sculpture in the United States: Joseph Beuys and the Work of Suzanne Lacy and Rick Lowe', *Public Art Dialogue*, 3.2 (2013): 144–167.

Kimmelman, Michael. 'In Houston, Art Is Where the Home Is', *The New York Times*, 17 December 2006, http://www.nytimes.com/2006/12/17/arts/design/17kimm.html.

---

58   Mouffe, 'Alfredo Jaar: The Artist as Organic Intellectual': 24.

Laclau, Ernesto and Chantal Mouffe, *Hegemony and Socialist Strategy: Towards a Radical Democratic Politics*, 2nd edition, London: Verso, 2014.

Lacy, Suzanne. 'Introduction', in Suzanne Lacy (ed) *Mapping the Terrain: New Genre Public Art*, Seattle: Bay Press, 1995, pp.19–47.

Marchevska Elena. 'Gómez-Peña and Balitronica Gomez, from La Pocha Nostra troupe', in Elena Marchevska, *the displaced & privilege: live art in the age of hostility*, London: LADA, 2017, p. 55-59.

Miranda, Carolina. 'How the Art of Social Practice is Changing the World, One Row House at a Time', *Artnews,* 7 April 2007, http://www.artnews.com/2014/04/07/art-of-social-practice-is-changing-the-world-one-row-house-at-a-time/.

Moore, Natalie. 'How Theaster Gates Is Revitalizing Chicago's South Side, One Vacant Building at a Time', *Smithsonian*, December 2015, http://www.smithsonianmag.com/innovation/theaster-gates-ingenuity-awards-chicago-180957203/.

Mouffe, Chantal. 'Artistic Activism and Agonistic Spaces', *Art & Research* 1.2 (2007): 1–8.

Mouffe, Chantal. 'Alfredo Jaar: The Artist as Organic Intellectual', *Revista Diseña*, 11 (July, 2017): 18–35.

Nance, Kevin. 'The Spirit and the Strength: A Profile of Toni Morrison', *Poets&Writers*, (November/December, 2008), https://www.pw.org/content/the_spirit_and_the_strength_a_profile_of_toni_morrison.

[On View], *Project Row Houses*, 2019, http://projectrowhouses.org/on-view.

Orlowski, Edward. 'House of Blues: The Shotgun and Scarcity Culture in the Mississippi Delta', in: Nnamdi Elleh (ed) *Reading the Architecture of the Underprivileged Classes: A Perspective on the Protests and Upheavals in our Cities*, London: Routledge, 2016, pp. 79-99.

Prince, Zenitha. 'Census Bureau: Higher Percentage of Black Children Live with Single Mothers', *Afro News*, 31 December 2016, http://www.afro.com/census-bureau-higher-percentage-black-children-live-single-mothers/.

Rabanera, April Garnet Sommer. *Success Stories: An Exploration of Three Non-profits Working in Preservation, Affordable Housing, and Community Revitalisation,* Master of Heritage Conservation, USC School of Architecture, Los Angeles, California, 2013.

Row House CDC, https://www.rowhousecdc.org/.

Sewing, Joy. 'Gentrification of the historic Third Ward', *Houston Chronicle*, 2 August 2018, https://www.houstonchronicle.com/business/real-estate/looped-In/article/Listen-Gentrification-of-the-historic-Third-Ward-13124478.php.

Shropshire, Corilyn. 'Chicago gentrification fears rise as East Garfield Par, Austin, South Lawndale housing prices increase', *Chicago Tribune*, 20 December 2018, https://www.chicagotribune.com/business/ct-biz-chicago-housing-costs-gentrification-worries-20181219-story.html.

Simms, Margaret Karina Fortuny, and Everett Henderson. 'Racial and Ethnic Disparities Among Low-Income Families', *The Urban Institute,* August 2009, https://www.urban.org/sites/default/files/publication/32976/411936-Racial-and-Ethnic-Disparities-Among-Low-Income-Families.PDF.

Smith, Stephanie. *The Art of Place and the Place of Art at Project Row Houses,* Master of Arts, Rice University, Houston, Texas, 1998.

Tucker, Jasmine and Caitlin Lowell. 'National Snapshot: Poverty Among Women & Families, 2015', *National Women's Law Centre,* September 2016, http://nwlc.org/wp-content/uploads/2016/09/Poverty-Snapshot-Factsheet-2016.pdf.

Tucker, Sheryl. 'Reinnovating the African-American Shotgun House', *Places* 10.1 (1995): 64–71.

Ukeles, Mierle. 'Maintenance Art Manifesto', in: Kristine Stiles and Peter Selz (eds) *Theories and Documents of Contemporary Art: A Sourcebook of Artists' Writings*, Berkeley: University of California Press, 1996 (1969), pp. 622–24.

Wainwright, Oliver. 'Chilean architect Alejandro Aravena wins 2016 Pritzker prize', *The Guardian*, 13 January 2016, http://www.theguardian.com/artanddesign/2016/jan/13/chilean-architect-alejandro-aravena-wins-2016-pritzker-prize.

[Young Mothers Residential Program], *Project Row Houses*, 2017, https://static1.squarespace.com/static/55832a9de4b0920b55b16891/t/588ba5c186e6c0ffe0e5d4c3/1508433636183/YMNRP+Application+2017.pdf.

Zilliacus, Ariana. 'Half a House Builds a Whole Community: Elemental's Controversial Social Housing', *Arch Daily*, 24 October 2016, http://www.archdaily.com/797779/half-a-house-builds-a-whole-community-elementals-controversial-social-housing.

COMPLICITIES, SOLIDARITIES AND EVERYTHING IN BETWEEN

Art and Political Engagement in the Housing Movement in Bucharest and Cluj

IOANA FLOREA & VEDA POPOVICI

The street is the art of small things. Write the microhistory of the street. The street is art. [...]
Criticize all authority! Be realistic! Demand the impossible![1]

What makes art projects become complicit with systems of power and what makes it possible for them to empower housing struggles, broadening their tactics, vision, and positionality? As active members in the Common Front for Housing Rights [FCDL] in Bucharest, we ask this question from engaged and participatory positions, exploring a variety of such intersections, with an analytical emphasis on the nature of ambivalence.

Art projects always walk the thin line between real solidarity on one side, and complicity, appropriation, museification/culturalization, facilitating gentrification and poverty porn on the other side. In this, there is a permanent negotiation, a tension between contradictions and potential. The causes of this ambivalence can be traced back to a variety of factors: independent artists' precarity and their need for alliances; the use of (tactical) artistic tools by persons affected by housing injustice; the possibility and need of funding/visibility for social causes (otherwise under-funded) through art projects. The causes of this ambivalence can also be found in the structural factors that affect housing conditions in Romania, art and artists' conditions, as well as possibilities and limits for housing struggles. All the actors involved, all their interactions ranging from solidarity building to conflict, and the results of those interactions, are framed within these structural factors.[2]

What characterizes these structural factors is a condition of semi-periphery for global capital: the territories occupied by today's Romania were always quite poor, predominantly rural, and penetrated by extractivist mechanisms exploiting labor and raw materials. Before WWII, most households could be found in situations of severe housing deprivation. Against this backdrop, the post-war regimes engaged in massive efforts to solve the housing crisis and, at the same time, to break the hegemony of the private property paradigm. These social as well as ideological issues were tackled with several policies starting in the late 1940s: nationalization of pre-1945 private properties belonging to owners with several properties; massive programs for building public housing blocks-of-flats; and the systematization and industrialization of rural areas.

---

1   From 'Manifesto of Active Art' in Maria Drăghici and Irina Gâdiuță (eds), *Reader Rahova Uranus Lum Doc 2009*, Bucharest: Vellant, 2010: 242. The manifesto is one of the many products that resulted from the collaboration between local communities and artists in the Rahova-Uranus neighborhood in Bucharest with the aim of documenting experiences of evictions and gentrification. See more in the section '*The Generosity Offensive*: artists bringing or resisting gentrification'.

2   Ioana Florea, Agnes Gagyi and Kerstin Jacobsson, *Contemporary Housing Struggles: A Structural Field of Contention Approach*, Palgrave, forthcoming 2021.

With the regime change after 1989, in Romania, as in other ECE countries, housing reform followed three main paths: the rapid and continuous sale of state-owned stock; deregulation of urban development, to the benefit of the private real-estate sector; and the re-privatization of the nationalized housing stock, through restitutions to pre-1950 owners, their heirs, or their legal-rights buyers. These policies were cemented by the influence of international financial institutions, such as the World Bank and IMF, overseeing the entire 'transition' process.[3]

Following long-term histories of structural racism,[4] the Roma population has been disproportionately targeted and affected by restitution-related evictions, blocked access to property and land rights, and restricted to areas underserved by public services and thus precarious housing.[5] The emergence of the housing movements in Romania is based on the fact that Roma emancipatory movements tackled housing (in)justice early on, in the 1990s, inspiring, supporting, and protecting neighborhood struggles for housing rights. The base for our housing movements also included the early *right to the city* coalitions and the struggles against restitution-related evictions from the city center of Bucharest, during the 2000s. In these struggles, intersections with the art scenes took place early on – as cross-class collaborations and solidarity, but also as complicities with the gentrifying capital.

Walking the Thin Line: Five Stories of Ambivalent Solidarity and Complicity

Strongly tied with cities, the art scenes that developed after 1989 have constantly addressed issues of public space, reflections on the commons, social engagement and critical reflections on locality.[6] Such interests can especially be found in the critical or alternative sections of the fields of contemporary art, theater, and dance.[7] Here, political theater, contemporary dance and contemporary art have intersected frequently and have resulted in social and professional alliances, and hybrid – sometimes activist – projects.

Art scenes have also made a constant material and critical effort to engage with space. Developing their own spaces and reflecting on the precarity and vulnerability of running such spaces has been a constant struggle and field of contestation for critical art scenes.[8] Often conflicted between the need for autonomy outside state institutions or market-driven interests, and the precarity that comes along with such efforts, art scenes, especially those based in Bucharest and Cluj (Romania's

3   Enikő Vincze, 'The Ideology of Economic Liberalism And The Politics Of Housing In Romania', *Studia Ubb. Europaea* LXII (2017): 29–54.
4   Gabor Fleck et al., *Come Closer. The Inclusion and Exclusion of Roma in Present Day Romanian Society*, The National Agency for Roma, 2008.
5   Enikő Vincze and George Zamfir, 'Racialized housing unevenness in Cluj-Napoca under capitalist redevelopment', *City Journal* 23.4-5 (2019): 439–460.
6   Such as the activities of the *e-cart* collective, see in Raluca Voinea, 'Public Space' in Tranzit Czechia, *Atlas of Transformation,* 2011. Available at: http://monumenttotransformation.org/atlas-of-transformation/html/p/public-space/public-space-raluca-voinea.html.
7   Iulia Popovici and Raluca Voinea, *Metaforă. Protest. Concept Performance Art din România și Moldova*, Idea Publishing, 2018.
8   Igor Mocanu, 'The paradoxical postcommunist utopia of artist-led spaces in Bucharest', in ArtychokTV (ed.) *Close-Up: Post-transition writings*, Editions of the Academy of Fine Arts in Prague, 2014, pp. 56-67.

two richest cities), have sought out and experimented with numerous forms of independent spaces – embedded in different forms of political negotiations, alliances, and conflicts.

Artists' needs for material resources – equipment, books, money, physical time and strength, spaces for work, for learning, for interaction, and for visibility – combined with the drive to tackle urban issues, have also produced difficult interactions between art scenes and social justice struggles. Embedded in the post-1990 hegemonic (economic, social, discursive) frames of anti-communism, aspiration towards the West/self-colonization as the corrupt East, capitalism and the unregulated market, many art projects and initiatives – such as the ones we discuss here – walk the thin line between cooptation and opposition. They pave the way for gentrification or help to resist it, they make social justice struggles visible or ignore them, they culturalize social struggles or up-scale/re-tool them.

1. The Generosity Offensive: Artists Bringing or Resisting Gentrification

In 2001, the property restitution law (the infamous Law 10/2001) was passed. The rising anti-communist discourse of the *transition* legitimated a turn towards property regimes similar to the pre-1945 one, dominated by large private properties and the unregulated land/real estate market. A new class of *restitution landlords*, partly overlapping with the pre-war landlord class, was on the rise, while state tenants living for decades in nationalized and then restituted buildings were affected by evictions, usually with no proper relocation, followed by the gentrification and re-development of entire neighborhoods.[9]

In the early 2000s, the impact of the restitution law became visible in Bucharest: entire areas in the pre-war central quarters were affected by evictions, thousands of people lost their homes, new developments and high-rises signaled the unfolding of gentrification processes. Rahova-Uranus was a working class ethnically-mixed neighborhood with Roma families, not far from the city center, with a tissue of pre-war individual houses, targeted by restitutions, gentrifiers, and developers especially during the pre-2008 boom. A few companies, with cultural labels such as the International Center of Contemporary Art [CIAC], DC Communication, Headvertising, The Ark cultural center, acquired buildings in the area and launched calls for community art projects to be coordinated by teams of young artists, in an attempt to speed up and artwash the gentrification process.[10] In 2006, *Ofensiva Generozității* [The Generosity Offensive, OG] project won such a call to implement artistic and educational activities with local dwellers – thus pacifying them before the evictions and turning their resistance into *generosity*.

But the local dwellers had thick networks among each other and a disco – LaBomba – where they would gather not only on the weekends, but also for spending time together among mothers,

9   Veda Popovici, 'Residences, restitutions and resistance: A radical housing movement's understanding of post-socialist property redistribution', *City Journal* 24.1-2 (2020): 97–111.
10   David Schwartz, 'Evacuați, artiști și agenți ai gentrificării: Proiectele din Rahova-Uranus / Interviu cu Cristina Eremia' [Evictees, artists and agents of gentrification: Projects from Rahova-Uranus / Interview with Cristina Eremia], *Gazeta de Artă Politică* [The Political Art Gazette], 2014, 5(2):6–7.

children, etc. Thus, the interaction between the young artists and the local social networks produced changes in both: the artists became increasingly involved in supporting local resistance against evictions, distancing themselves from the gentrifiers by the end of 2007. On the other hand, the dwellers became increasingly involved with allies from the art scene and started using artistic tools to gain public visibility for their struggle. Together they organized music, theater and drawing workshops for children in the area; they produced several documentaries and video installations, an exhibition called *Fake Documents* revealing the illegal operations of *Cherecheş & Cherecheş* lawyer office (a key actor in Bucharest's restitutions); held film screenings, shows, and concerts for a wide audience – which provided further allies to the local resistance. With this renewed strength, protest actions and confrontational meetings with local authorities – accomplices in the gentrification process – were also organized together.

In 2009, LaBomba disco was turned into a community center, a base for community organizing, for alliance building between artists and locals, for organizing against the illegal restitutions and evictions taking place in the area. This illustrates the coming together of resources and needs of both the local activists – strongly represented by women and youth – and the involved artists.[11] LaBomba was evicted in 2011, through an illegal restitution process opposed by local residents. The solidarity around it was unprecedented, also leading to the conception of the critical theater play *Fără sprijin* [Without support, 2012], revealing the fruitfulness of almost five years of alliance-building and solidarity among artists and the Rahova-Uranus dwellers.

Fig. 1: The eviction of LaBomba Cultural Center, 2011. Photo by Maria Drăghici.

---

11   Maria Drăghici and Irina Gâdiuţă (eds), *Reader Rahova Uranus Lum Doc 2009*, Bucharest: Vellant, 2010. Available at: https://issuu.com/maria_draghici/docs/labombastudios.blogspot.com.

In 2013, one more family was evicted, following yet another restitution: the head of the evicted family was one of the LaBomba community center organizers.[12] A massive protest was organized in solidarity with her in spring 2013 and the first Forum for Housing Justice a few months later, gathering activists from Bucharest and Cluj city, hosted by the Tranzit.ro/Bucharest art space. Towards the end of 2013, the strength gathered through these collective actions (and the general anti-austerity fury of those years) was galvanized into setting up the Common Front for Housing Rights as a cross-class activist platform clearly positioning itself against evictions, for the support of evictees and for the right to adequate housing for all. The artistic projects that followed in alliance with FCDL had a clear political message, as we will see in the next section.

Thus, what began in 2006 as a community art project, on the thin line between gentrification and empowerment, produced the conditions for an enduring housing movement that is still active today and continues producing new alliances.

2. Evicting the Ghost: Evictions Happening Here and Afar

In 2010, just as the Rahova-Uranus community was beginning to organize more and more strongly against evictions while experimenting with educational and artistic tactics, Bucharest's contemporary art scene welcomed the launch of the book *Evicting the Ghost. Architectures of Survival*.[13] Produced by a mixed team of artists, architects, and sociologists, the volume is an analysis of the shelters and shacks created by evictees after losing their homes to restitutions.

The years leading up to the launch were arguably the period when such improvised shelters of evictees had been the most visible in the city. Scattered generally around the city center and of temporary nature, these shelters and makeshift shacks have been used as both housing and means of protest by the evictees.

The volume follows the processes of restitution and traces its historical context, including information on the legislative history and a series of case studies. Its starting point is an architect's amazement in encountering a makeshift shack of an evicted family while randomly walking the streets. The book's approach thus established a clear-cut distance between the architect/artist/researcher and the evictees. As the editor puts it: 'a distance specific to a scientific atlas'.

The makeshift shacks became the central motif of the volume comprising several researchers' contributions and architect illustrations. The homes themselves were looked at with a romantic gaze, considered as *ghost-homes*. Although including direct accounts of evictees, there was no analysis on the implications of gentrification, neo-liberal urban transformations, or the hegemony of private property. Nor was there any recognition of the political character of the

12   Mihaela Michailov and David Schwartz, 'Înainte eram puțini, acum suntem mai mulți și o să fim și mai mulți!' [Before there were less of us, now we're more and more will come!], *Gazeta de Artă Politică* [The Political Art Gazette], 2013. Available at: http://artapolitica.ro/2013/03/11/inainte-eram-putini-acum-suntem-mai-multi-si-o-sa-fim-si-mai-multi.

13   Alina Șerban (ed.) *Evicting the Ghost. Architectures of Survival*, Bucharest: Centrul de Introspecție Vizuală, 2010.

shacks as a protest tactic. In the vein of a *scientific* project, the case studies became 'dioramas illustrating the conflicts of the post-socialist Romanian society'.

The political process of evictions was used as an object to be analyzed in its own staging, in the *décor* of the contemporary city. Rather than material signs of violence produced by the capitalist enclosure of resources in post-socialism, the shacks were considered to form 'a typology of the dwelling places developed by those evicted, expressions of nowadays urban survival'.

In conclusion, the project can be situated in an ambivalent position in which the distance established by the authors' gaze guarantees a dis-engaged relation towards the evictees. However, in the concrete context of 2010 Bucharest, *Evicting the Ghost* made visible a violent process by signaling the extreme housing conditions faced by evictees[14] and thus, could contribute to opening the space for a wider, more critical conversation on restitutions and evictions.

3. Train Delivery: Enter the Creative Class

While *Evicting the Ghost* represents a type of culturalizing political processes, by taking the position of a so called *objective* researcher, the 2013 project *Train Delivery* represents a different approach, one in which the creative class takes center-stage and reclaims the city for itself.

Train Delivery (the original title is in English), a festival 'about creative people reinventing the Northern Train Station' in Bucharest is part of the franchise festival *Street Delivery*, produced by *Cărturești* – the most successful bookshop-chain in Romania. The main train station in Bucharest, a busy and socially tensed area, is identified as the place where 'most have begun their love or hate story with Bucharest'. 'Love and hate' structure the whole program of events, identifying an area of the city that needs an intervention from the *creatives*. Their *reinvention* of the neighborhood is a vision of arts & crafts workshops, *creative* commodities entrepreneurship, and debates around the historical heritage of the area. A certain feeling of experimentation, innovation, and avant-garde permeates the events that seem to be part of a totally different world than the local one.

From the outset, the event stated that there was free entrance for 'all those who love their city'. Talking about love and hate for the city in the context of the train station neighborhood is at best cynical. Being quite close to the city center, the area is a vital spatial resource for its many residents from highly precarious communities of homeless people, sex workers, and drug users; and, thus, it is widely considered unsafe by the middle class. The festival becomes an apparatus meant to correct such an impression and make the area appealing (again) for such middle class.

One event in particular marks the distance between the creative visitors and the locals: *Train Delivery* includes a workshop organized by the Police titled *How to be safe from sneaky*

14   Daniela Calciu, 'Alex Axinte and Cristi Borcan (Studio BASAR), Evicting the Ghost. Architectures of
      Survival, Bucharest: Center for Visual Introspection', *Colloquia. Journal for Central European History*,
      Babes-Bolyai University Cluj, Vol. XIX (2012): 247–249.

thieves wherever you are. Obviously meant for the visitors of the festival and not catering to the locals, the workshop draws a thick line of criminalization between the creative class and the neighborhood's communities. An expression of the creative class's complicity with repressive policies, the event marks an unambiguous positioning on the side of power and authority.

The workshop with the police is just one of the indicators of the festival's consistency in affirming hegemonic ideologies and policies. Another dimension is found in its aspirational eurocentric perspective: the area of the train station is to 'rebecome a civilizing space', an objective attained through 'simple creative means'. The audience is addressed as 'citizens' that want to 'reappropriate the train station as a cultural good that generates a state of well-being'. The civilizing 'reinvention' of the area is done with the aim of 'reaffirming its European spirit'.

The civilizing eurocentric components do not remain in the discursive realm. Rather, they mark the coherence with broader urban transformations in that part of the city. Situated midway between the central government's headquarters and the parliament [House of the People], the train station area has long caught the attention of private real estate developers. All of the discursive indicators described above amount to a clear class threshold between the festival's producers and audience (part of the creative class) and the local residents. In this context, through the means of art practices and discourses of creativity, artists are co-opted through a directly complicit relationship with power and an indirectly complicit relationship with private real estate interests.[15]

4. Carol 53: Squatting Versus Anti-squatting

Artwashing of the violence in the city can take different forms: such as art students nicely painting a fence that segregates a neighborhood with predominantly Roma families in Baia Mare town, invited by the mayor who then tries to classify the fence as an object with heritage value; or music and art festivals for the better-off organized in poor areas, paving the way for gentrification and related evictions (for example in Rahova, Ferentari, Mătăsari neighborhoods, or in the train station in Bucharest). But artists can also take part directly, as engaged actors in the processes of urban violence. That direct engagement can range from graffiti painters complaining about homeless persons searching for shelter in the same abandoned buildings targeted by graffers, to taking part in actual evictions – in the context of the harsh competition over space.

In the summer of 2012, a group of young artists and architects in Bucharest started restoring an old building with heritage value, transforming it into an independent cultural center – Carol 53. In the process, they evicted a Roma family with children, previously living in the house informally. During public presentations of their project, the artists would admit to the eviction saying that cohabitation was not possible. The manager of the building – himself a respected architect

---

15   Veda Popovici, 'Delivering the City into the Arms of Capital: Gentrification and the Harmlessness of Art in Bucharest', *Gazeta de Artă Politică* [The Political Art Gazette], 2014. Available at: http://artapolitica.ro/en/2014/04/27/delivering-the-city-into-the-arms-of-capital-gentrification-and-the-harmlessness-of-art-in-bucharest/.

and participant in the movement for the protection of pre-war buildings with 'heritage value' – supported the Carol 53 artistic collective. This reflected the dominant view in the heritage protection movement that poor dwellers are destroying heritage value, while those buildings are better suited for cultural and entrepreneurial activities than housing (of modest means).

Fig. 2: Arnold Schlachter, Carol 53 house, graphite on paper.

Critical reactions were nonetheless formulated by several people and groups, arguing against the inequalities embedded in current urban development models and against this act of *squatting the squatters* – possibly becoming a new social cleansing practice invested with cultural value.[16] Researchers and activists continued to criticize the multiplication of the 'Carol 53 model' of gentrification, which was gaining ground in other cities such as Cluj and Timişoara. Young artists and architects continued their events and parties, legitimizing their initiative as an innovative one in the context of Bucharest, calling it the first 'cultural squat' in the city. Despite

16   Roxana Bucată, 'Artişti şi homeleşi. Care pe care squatează' [Artists and the homeless. Who's squatting who], *Think Outside the Box*, 1 August 2012, http://totb.ro/artisti-si-homelesi-care-pe-care-squateaza; Mihai Codreanu, Arnold Schlachter and Veda Popovici, 'Guarding the ruins – two episodes on gentrification and anti-squatting'. *Gazeta de Artă Politică* [The Political Art Gazette], 2014. Available at: http://artapolitica.ro/en/2014/06/05/guarding-the-ruins-two-episodes-on-gentrification-and-anti-squatting/.

being established with the building owner's permission, Carol 53 was self-branded as a 'squat' for its self-conscious organization/cohabitation – trying to differentiate itself from the informal occupations determined by deprivation and from its previous dwellers. Branded as a squat, Carol 53 positioned itself as a promoter of a desirable (at the same time alternative and yuppie) culture, with extra glow from Western currents of the *squatting culture*, while managing to avoid taking responsibility for the eviction it had caused, and avoiding taking a position towards housing inequalities and violent homelessness in the city.

Carol 53, with its supporters from the art scene and from the heritage protection movement, not only evicted the informal dwellers, but also denied their identity of *squatters*, trying to confine them as 'persons without documents', to dismiss them as 'illegal occupants', destroyers of the building, ignorant to its cultural value.[17] Through this labeling, they were also trying to deny the previous dwellers' symbolic association to global housing struggles and squatting phenomena.

5. The European Capital of Culture: the Periphery (Un)Becomes Western

The art scenes in Romania, as in other (semi)peripheral contexts, look towards the West from a subaltern gaze[18] – not only in terms of artistic contents and collaborations, but also in terms of desired and needed material support. The European Capital of Culture [ECoC] program is very revealing in terms of how such material support is organized and awarded, and how local art scenes respond to it – as it is the main program through which the EU intervenes in national/local cultural policies since the 1980s.[19]

Since the 1990s, the ECoC title is awarded to smaller or less affluent or less known cities with growth potential, which are ready to use culture as a developmental path. Thus, in competing for this title, cities must promise development and must create fantasy visions of a bright future[20] – with the necessary ingredients and catch-words, such as europeanness, freedom of expression, tourism, new grandiose cultural buildings, cultural entrepreneurship, diversity. In such visions, the cities become dreamscapes and playgrounds; and the artists and cultural managers prepare them for about five years in advance of the planned ECoC festival – hoping that it will boost their well-being, which hardly ever happens.[21] In 2007, the title was awarded to the Romanian city of Sibiu. It was a welcomed occasion for several independent artists to access funding and wide public exposure. It was also a strong gentrifying mechanism: the racist, classist, ageist,

17 Ioana Florea and Mihail Dumitriu, 'Transformations of housing provision in Romania: Organizations of subtle violence', *LeftEast*, 24 October 2018, https://lefteast.org/transformations-of-housing-provision-in-romania-organizations-of-subtle-violence.
18 Veda Popovici and Ovidiu Pop, 'From over here, in the periphery: a decolonial method for Romanian cultural and political discourses', *LeftEast*, 11 February 2016, https://lefteast.org/from-over-here-in-the-periphery-a-decolonial-method-for-romania.
19 Alexandra Oancă, *Governing the European Capital of Culture and Urban Regimes in Sibiu*, Master thesis, Central European University, Budapest/Hungary, 2010.
20 Alexandra Oancă, *Bidding Wars: Enactments of Expertise and Emotional Labor in the Spanish Competition for the ECoC 2016 Title*, PhD diss., Central European University, Budapest/Hungary, 2018.
21 Philip Boland, '"Capital of Culture – you must be having a laugh!" Challenging the official rhetoric of Liverpool as the 2008 European cultural capital', *Social & Cultural Geography* 11 (2010): 627–645.

social cleansing of its city center was hardly acknowledged, yet had a huge impact on urban morphologies.[22]

Again in 2016, several cities in Romania embarked on the ECoC competition, and city councils spent considerable public budgets on foreign experts and consultants for conceiving applications. Four cities were shortlisted: Bucharest, Cluj, Timișoara, and Baia Mare – despite their well-known fierce eviction practices.[23] Many independent artists participated voluntarily in conceiving these applications, hoping to access (much needed but still quite limited) funding and support once the competition was won. Very few art collectives refused to collaborate on such endeavors. Many were co-opted and, from this position, proposed small projects more or less involving marginalized groups: projects of art education, documenting histories of such groups and turning them into artistic objects, interventions in less affluent neighborhoods. As expected, participants from such neighborhoods were never involved as paid staff. Art, education, and community participation were used as magic concepts that would solve everything in the candidate cities,[24] thus rendering harsh social problems of inequality and dispossession insignificant. Urban redevelopment projects (proposed by local authorities together with real estate companies) legitimized by cultural projects/buildings were not challenged or opposed, despite their clear gentrifying character. The ECoC urban fantasies, together with the precarity of the artists, proved to be very effective anesthetics.

Timișoara won the competition in 2016, becoming the ECoC in 2021, along with Novi Sad in Serbia and Elefsina in Greece. Evictions had already started before 2019, in central areas planned for redevelopment. Homeless persons are violently removed from areas where they could be visible to tourists and aspiring middle classes.

A critical group formed around the Political Art Gazette [GAP], and followed the submissions of the four shortlisted cities, revealing the EU control mechanisms, their complicity in processes of uneven development and local dispossession. GAP is an independent activist online and print magazine launched in 2013 that 'discusses, analyzes and promotes the social and political dimension of the most diverse forms of cultural and artistic projects', following the motto 'all art is political'.

GAP has always been present as part of the housing struggle, having interconnected roots with LaBomba collective and the Popular School of Contemporary Art [ȘPAC, of which you will hear more about in the next section], and offering a platform for creating/disseminating discourse, amplifying voices, building up alliances between the art scenes and the housing movement. GAP15 (2016) was a special issue dedicated to investigating the ECoC program and its economic mechanisms legitimized by cultural discourses and dreamscapes. It featured social research contributions from previous ECoC contexts and analysis of local urban and art contexts in which

---

22 Oancă, 2010.
23 Bucharest registered tens of thousands of forced evictions; Cluj has more than 1.000 persons evicted on the city's landfill Pata Rât; Timișoara evicts homeless persons even from the empty fields at its periphery; Baia Mare has the infamous fence separating a neighborhood with predominantly Roma families, and relocated evictees in unsafe formerly industrial areas.
24 George Zamfir, 'Cluj Capitală Culturală sau justiție culturală ca soluție la injustiții sociale?', *Gazeta de Artă Politică* [The Political Art Gazette], 15 (2016):12–13.

the ECoC would plant itself and pick up the yields for private profits. It raised otherwise silenced questions and dissatisfactions, and also invited other cultural actors to participate in this critical effort. Thus, beyond the ECoC frame of cooption, there is also potential for opposition.

Holding the Line: Three Stories About Art as Political Tool Within the Housing Movement

1. Cluj: Integrating Art Tactics within Activist Strategy

Cluj-Napoca, one of the largest Romanian cities and the first in terms of housing costs (rents and sales), has emerged as the *Silicon Valley of Europe* with its boom of the IT industry. Welcomed by both the local and central government, this economic shift has rapidly transformed the city into the main candidate to be a *European city*: gentrified, with a growing population of tenants, and highly appealing to unicorns such as Uber or AirBnB. Quickly augmenting housing prices, hurried evictions, and the targeting of Roma residents considered *undesirable* have marked the city for more than a decade.

In parallel, a radical political scene is emerging, organizing struggles and forging critical analysis against these processes. One of the most important events to mark the development of this activist scene was the eviction of over 300 people from Coastei street on the early morning of 17 December 2010. Forced by a high number of police officers pressing for the eviction, the community was dislocated to the city margins, to what they soon found to be container substandard housing close to the city's main landfill and chemical waste dump, known as Pata Rât. The eviction displaced the entire community, including many families of Roma ethnicity, to an area cut off from public transport, health services, and access to schools.

Organizing around this struggle began immediately after the eviction. The community found support in local activists, artists, and academics and in NGOs supporting Roma rights. Soon, the alliance made the struggle considerably visible and marked the beginning of the housing movement in Cluj. The activist community around Pata Rât developed tactics suitable for grasping the attention of the dynamic, aspirational city of Cluj. They chose a variety of artistic activist means to draw attention to the Pata Rât injustice and gain more supporters.

According to Enikő Vincze, one of the struggle's main organizers and activist researchers, the alliances created with the local critical art scene proved to be effective, innovative and empowering, establishing an original type of *artistic activism*.[25] One of the most visible artistic interventions was the project organized by Şcoala Populară de Artă Contemporană [Popular School of Contemporary Art – ŞPAC] in 2011: an educational and activist endeavor, *Eşti în Pata Rât* [You are in Pata Rât] produced several art interventions made by students in solidarity with the evictees. Exhibited as art shows and used as props for *welcoming* officials in protest actions, the works used the metaphoric, ironic, and intellectual style of contemporary art to draw attention and raise awareness towards the

---

25   Mihaela Michailov, 'Am pus împreună bazele unui concept de activism artistic', *Gazeta de Artă Politică* [The Political Art Gazette], 2014. Available at: http://artapolitica.ro/2014/05/13/am-pus-impreuna-bazele-unui-concept-de-activism-artistic/.

emergency and gravity of the evicted community's situation.[26] Together with activist performances integrated in protests, the art interventions became one of the trademarks of the activist network formed around the Pata Rât/Coastei struggle.

Fig. 3: Making of Memorial Plaque by Dénes Miklósi, 2015. Photo credits: Noémi Magyari. Text reads: 'In memory of the 300 people, the majority of whom citizens of Roma ethnicity, forcefully evicted on 17 December 2010 by local authorities from Coastei street to an area next to Cluj's landfill'.

Later that year, the documentary *Prea-fericiţii din groapa de gunoi* [The Blessed of the Landfill] was launched. Produced by Solitude Project, the film was made by a multidisciplinary artist collective. Among the authors were choreographer Mihai Mihalcea, scriptwriter Michaela Michailov, and actress Katia Pascariu, all contemporary art producers with an engaged track, associated with the previously mentioned Generosity Offensive [OG] and the artist-run space subRahova. The documentary, while representing in a dignified manner the struggle and hardships of the Pata Rât community, reveals the involvement of the Orthodox Church in the affair: the Orthodox Church received the restitution of the land on Coastei street, and built a dorm for Theology University students. Shown in several cities, the documentary revealed the church as the main beneficiary of the eviction: showing who and how profited from the evictions significantly contributed to the framing of the Pata Rât struggle in broader economic phenomena, effectively politicizing it.

Producing a variety of tools and tactics inspired by artistic media, the collaboration between artists and activists laid the grounds for the wide visibility of the Pata Rât struggle, one of the

---

26   ŞPAC-Şcoala Populară de Artă Contemporană/Protokoll, 'Eşti în Pata Rât', in Adrian Dohotaru, Hajnalka Harbula and Enikő Vincze (eds) *Pata*, Cluj: Efes, 2016, pp. 138–162.

main catalysts of the radical housing movement in Romania. As we write, the Cluj art spaces tranzit.ro and Casa Tranzit remain important venues for the local housing movement generally associated with the Căşi Sociale organization [Social Housing NOW] and Asociaţia Chiriaşilor Cluj [ACC, Tenants Union Cluj].

2. Bucharest: Theater and Performativity as one more Political Tool

If Cluj developed a method of integrating artistic expressions in wider activist tactics, with an emphasis on object and performance as media, the typical development of the intersection between art and the housing movement in Bucharest consists of theater work. This can be traced back to the projects of the Rahova-Uranus community and OG collective. As early as 2008, the collective active in the Rahova-Uranus neighborhood, comprising both artists and local community members, produced the short theater play *Afară!* [Get Out!]. A piece based on the experiences of evictees and residents at risk of eviction, the play was meant to be a reflection on the hardships surrounding an eviction. However, the play would come to be considered the stepping stone for future alliances was *Fără Sprijin* [Without Support, 2012]. The play had several components that would recur in other plays documenting and illustrating the experiences of evicted communities: artists and community members co-producing and co-acting, the centrality of the evictees' experiences, and the emphasis on women's voices and perspectives. Featuring three of the main organizers and community members, Cristina Eremia, Cornelia Ioniţă, and Gabriela Dumitru, the play set the tone for future productions and the use of theater plays as a means for strengthening communities of action.

Fig. 4: LaBomba Studios, No Support theater play, poster.

In 2014, the play *La Harneală* [Wisecracking] was produced by the feminist Roma theater company Giuvlipen. The piece explores the phenomenon of evictions through the lens of evicted people, especially women and children, with a focus on the relationship with the public authorities. Tackling issues of racism, appropriation, and exoticization, the play marks a sophisticated critical reflection on the limits of housing activist organizing. Two years later, the *Subjective Museum of Housing* was first staged in the autonomous cooperative space *Macaz* and then in other locations throughout the country. The play ironically critiques the museification of the political experience, by adopting the format of a guided tour through a museum. In the mock tour, the actresses restage episodes from their own lives meant to portray the violence, injustice, pain, and solidarity of going through evictions. By showcasing their experiences in this way, the production team points out the constant importance and urgency of communicating the experiences of evictions to a wider audience, while making a bitter commentary on Bucharest's short memory on the issue.

Fig. 5: Scene from The Subjective Museum of Housing, 2018. Photo by HomeFest.

To this series of plays we can also add *Domiciliu Instabil* [Unstable Home] from 2016. Produced by a political theater collective consistently involved in the housing movement, the play explores the life experiences of the residents of the Moses Rosen senior home in Bucharest.[27] As most of the staged experiences belong to Jewish seniors, the play portrays a violent and intense

---

27   Gazeta de Artă Politică [The Political Art Gazette], 'A subjective history of housing in Romania as seen by the Moses Rosen rest home residents. The interwar period, 1920 – 1930 (part I and II)', *Gazeta de Artă Politică*, 2016. Available at: http://artapolitica.ro/en/2016/10/08/romana-o-istorie-subiectiva-a-locuirii-in-romania-din-perspectiva-rezidentelor-si-rezidentilor-caminului-moses-rosen-perioada-interbelica-anii-1920-1930/ and http://artapolitica.ro/en/2016/10/09/romana-a-subjective-history-of-housing-in-romania-as-seen-by-the-moses-rosen-rest-home-residents-the-interwar-period-1920-1930-part-ii/.

history of housing changes through fascism, socialism, and capitalist post-socialism. With the unique long-term perspective of the residents, the play adds another dimension to the position of the local housing movement – incorporating long historical perspectives into its political position. Community organizing and resistance against abusive and repressive measures, such as evictions from homes, become even more legitimate by adding this historical perspective.

The majority of the roles in all these plays are interpreted by actual community members. In the first three examples, these are women at the forefront of the housing movement in Bucharest.[28] Produced with the explicit scope of agitating the audience towards joining the movement, the plays are expressions of the methodologies developed by several political theater groups from Bucharest.[29] Both informing and empowering, the political theater methodologies are meant to boost local political struggles and to broaden their scope, tactics, and vocabulary. For the housing movement, the plays created a material, discursive, and affective space in which the experiences of evicted women can be expressed on their own terms. By creating infrastructures in which community members can interact, the plays constitute a vital space of bonding and collective cohesion. Marked by self-reflection and critical commentary, the plays build a dynamic space of debate in which key questions can be asked: what is the relationship between affected people and people in solidarity? What makes one an activist? What makes one an artist?

3. 'History Does (Not) Repeat Itself' in Cluj and Bucharest

One recent contemporary art project that draws from and feeds back into the housing movement is Veda Popovici and Mircea Nicolae's *History Does (Not) Repeat Itself*. Produced in 2017—2018, the project develops ten entangled political histories in post-socialist Romania that sit on the edge between fact and fiction.[30] The project, comprised of two exhibitions, several art pieces, a film, and several workshops and screenings, builds on a speculative history methodology of collective introspection and production. Through several workshops, the authors engaged with groups of activists, researchers, and artists to excavate fictional political stories of the post-89 period in Romania based on real events. The selection and editing process of these stories resulted in a series of political dreams and utopias that the participants feel particularly invested in. Speculating on what could have happened, the project reveals how close real people actually were to taking a different historical path for their political needs and desires. The subsequent exhibitions and film are composed of concrete artifacts that stand as testaments to these histories: protest banners, armbands, press releases, mock-up websites, posters, and building models, etc.

Two out of the total ten speculative stories build on the housing movement's experience. One of them, developed with activists and artists from Cluj, draws upon the infamous Coastei eviction that resulted in the Pata Rât settlement and ultimately ended up with the Theology dorm built on

---

28   Mihaela Michailov, 'Casa noastră cea de unele zile', *Gazeta de Artă Politică* [The Political Art Gazette], 22 January 2015, http://artapolitica.ro/2015/01/22/casa-noastra-cea-de-unele-zile-la-harneala.

29   Mihaela Michailov and David Schwartz, *Teatru Politic* [Political Theater], Bucharest: Tact, 2017.

30   Veda Popovici and Mircea Nicolae, *Istorii post-revoluționare* [Post-revolutionary histories], Project publication, 2018.

the land from which the Coastei community was evicted. But in the speculative history, the dorm itself, a real housing complex belonging to the Orthodox Church, becomes a site of resistance. The story unfolds as a tale of radical solidarity in which the students housed in the dorm learn about the foundational injustice of their building and open up the dorm to welcome anyone in need for shelter after an eviction. Their resistance goes against the politics of the Theology School or the Orthodox Church and instead joins forces with the local radical housing movement. As a proof of the possibility of such a story, the group in Cluj produced a stencil and banner that illustrate the moment in which the students housed in the dorm open the building up for evictees.

Fig. 6.:eti Pataki, 'The Theology Dorm Opens for Evictees', stencil, part of History Does (Not) Repeat Itself.

The other story is based on the existing residential complex Rose Garden in Bucharest, a massive housing project numbering about 900 apartments. It is built on the land of a former (privatized and dismantled) socialist factory, by the real-estate broker and developer Coldwell Banker – a giant of the real-estate sector originating from San Francisco. During a workshop organized in Bucharest, this capitalist story of privatization, gentrification, and extraction was turned into a story of workers' empowerment: the former factory, processing basic fabrics, is privatized in the late 1990s not to some foreign capitalist venture, but to its own workers becoming a collectively owned company. Then, it was the workers who decided to move the factory to smaller grounds, demolish the old factory and build a vast housing complex for themselves on this land. This way, the factory was saved from bankruptcy and extended in a vast community-owned structure including land, processing facilities, and housing units. The housing complex ended up being named *White Flower*, celebrating the cotton plant, the main fabric processed in the factory.

These two speculative histories paint two sides of the same story of the privatization of housing and the historical delegitimization, dislocation, and dispossession of workers in post-89 Romania.

While building on real events and affects, they also reveal the distance between utopia and reality and blur the lines between past and future. Bridging the gap between what has been done and what could be done, the whole project becomes a tool for dreaming a radical political future and mapping possible plans of action.

Reading Between the Lines

Writing from the perspective of participants and producers involved in the housing movement and art scenes, we sought to map some important events, projects and developments in Bucharest and Cluj that marked the features of the post-socialist critical/engaged art scenes, the complexity and strength of the local housing movement, and the tensions in these fields that separate solidarity practices from capital-complicit ones. With a keen eye on the relations of power they express, we propose a critique of the complicities with hegemonic neoliberal discourses and policies, while also tracing the intricate motivations of some of the actors involved. Trying to avoid relativizing, we explore the ambivalences that art projects may embed – walking the thin line between enabling and resisting gentrification, between dis-engaging with evictions and making them visible, between complicit relationships with power or private real estate interests and resisting the subaltern role of the artist in a (semi)peripheral context. We consider them as episodes in a broader history of art versus power.

Building upon our positive experiences and the inspiring work of our comrades – integrating art tactics within activist strategies, using theater and performativity as political tools, celebrating possible pasts and political dreams – we propose the housing movement's intersection with art as somewhat of a story of fulfilled solidarity possibilities: they do reinforce each other in radical and empowering ways. We have seen it happen. Although we do not think there is any strict recipe for how to do art and empower social movements, or how to do political work and mold the imagination and affects of comrades and peers, we do consider the stories told here to be valuable insights and, to some degree, lessons for our radical futures.

References

Boland, Philip. '"Capital of Culture – you must be having a laugh!" Challenging the official rhetoric of Liverpool as the 2008 European cultural capital', *Social & Cultural Geography* 11 (2010): 627–645.

Bucată, Roxana. 'Artiști și homeleși. Care pe care squatează' [Artists and the homeless. Who's squatting who]. *Think Outside the Box,* 2012. Available at: http://totb.ro/artisti-si-homelesi-carepe-care-squateaza/.

Calciu, Daniela. 'Axinte, Alex and Borcan, Cristi (Studio BASAR). Evicting the Ghost. Architectures of Survival, Bucharest: center for Visual Introspection', *Colloquia. Journal for Central European History*, Babes-Bolyai University Cluj, Vol. XIX, 2012, pp. 247-249.

Capitale ale Culturii în contextul politicilor UE [Capitals of Culture in the context of EU politics], Gazeta de Artă Politică, issue 15 (4), 2016.

Codreanu, Mihai, Schlachter, Arnold and Popovici, Veda. 'Guarding the ruins – two episodes on gentrification and anti-squatting', *Gazeta de Artă Politică*, 2014. Available at: http://artapolitica.ro/en/2014/06/05/guarding-the-ruins-two-episodes-on-gentrification-and-anti-squatting/.

Drăghici Maria and Gâdiuță, Irina (eds), *Reader Rahova Uranus Lum Doc 2009*, Bucharest: Vellant, 2010.

Fleck, Gabor et al., *Come Closer. The Inclusion and Exclusion of Roma in Present Day Romanian Society*, The National Agency for Roma, 2008.

Florea, Ioana and Dumitriu, Mihail. 'Living on the edge: the ambiguities of squatting and urban development in Bucharest', in Freia Anders and Alexander Sedlmaier (eds) *Public Goods versus Economic Interests. Global Perspectives on the History of Squatting*, New York and London: Routledge, 2017, pp. 188–210.

_____. 'Transformations of housing provision in Romania: organizations of subtle violence', *LeftEast*, 24 October 2018, https://lefteast.org/transformations-of-housing-provision-in-romania-organizations-of-subtle-violence.

Florea, Ioana, Gagyi, Agnes and Jacobsson, Kerstin. *Contemporary Housing Struggles: A Structural Field of Contention Approach*, Palgrave, forthcoming 2021.

Gazeta de Artă Politică, 'A subjective history of housing in Romania as seen by the Moses Rosen rest home residents. the interwar period, 1920 – 1930 (part i and ii)', *Gazeta de Artă Politică*, 2016. Available at: http://artapolitica.ro/en/2016/10/08/romana-o-istorie-subiectiva-a-locuirii-in-romania-din-perspectiva-rezidentelor-si-rezidentilor-caminului-moses-rosen-perioada-interbelica-anii-1920-1930/ and http://artapolitica.ro/en/2016/10/09/romana-a-subjective-history-of-housing-in-romania-as-seen-by-the-moses-rosen-rest-home-residents-the-interwar-period-1920-1930-part-ii/.

Michailov, Mihaela and Schwartz, David. 'Înainte eram puțini, acum suntem mai mulți și o să fim și mai mulți!' [Before there were less of us, now we're more and more will come!]. *Gazeta de Artă Politică*, 2013. Available at: http://artapolitica.ro/2013/03/11/inainte-eram-putini-acum-suntem-mai-multi-si-o-sa-fim-si-mai-multi/.

Michailov, Mihaela. 'Am pus împreună bazele unui concept de activism artistic', *Gazeta de Artă Politică*, 13 May 2014. Available at: http://artapolitica.ro/2014/05/13/am-pus-impreuna-bazele-unui-concept-de-activism-artistic/.

_____. 'Casa noastră cea de unele zile' [Home sweet home]. *Gazeta de Artă Politică*, 2015. Available at: http://artapolitica.ro/2015/01/22/casa-noastra-cea-de-unele-zile-la-harneala.

Michailov, Mihaela and Schwartz, David. *Teatru Politic* [Political Theater], Bucharest: Tact, 2017.

Mocanu, Igor. 'The paradoxical postcommunist utopia of artist-led spaces in Bucharest', in ArtychokTV (ed.) *Close-Up: Post-transition writings*, Editions of the Academy of Fine Arts in Prague, 2014, pp. 56-67.

Oancă, Alexandra. *Governing the European Capital of Culture and Urban Regimes in Sibiu*, Master thesis, Central European University, Budapest/Hungary, 2010.

_____. *Bidding Wars: Enactments of Expertise and Emotional Labor in the Spanish Competition for the ECoC 2016 Title*, PhD diss., Central European University, Budapest/Hungary, 2018.

Popovici, Veda. 'Delivering the City into the Arms of Capital: Gentrification and the Harmlessness of Art in Bucharest', *Gazeta de Artă Politică*, 27 April 2014, http://artapolitica.ro/en/2014/04/27/delivering-the-city-into-the-arms-of-capital-gentrification-and-the-harmlessness-of-art-in-bucharest/.

Popovici, Veda and Pop, Ovidiu. 'From over here, in the periphery: a decolonial method for Romanian cultural and political discourses', *LeftEast*, 11 February 2016, https://lefteast.org/from-over-here-in-the-periphery-a-decolonial-method-for-romania.

Popovici, Veda and Nicolae, Mircea. *Istorii post-revoluționare (Post-revolutionary histories)*, Project publication, 2018.

Popovici, Veda. 'Residences, restitutions and resistance: A radical housing movement's understanding of post-socialist property redistribution'. *City Journal* 24.1-2 (2020): 97-111.

Popovici, Iulia and Voinea, Raluca. *Metaforă. Protest. Concept Performance Art din România şi Moldova* [Metaphor. Protest. Concept Performance Art in Romania and Moldova], Idea Publishing, 2018.

Schwartz, David. 'Evacuaţi, artişti şi agenţi ai gentrificării: Proiectele din Rahova-Uranus / Interviu cu Cristina Eremia' [Evictees, artists and agents of gentrification: Projects from Rahova-Uranus / Interview with Cristina Eremia], *Gazeta de Artă Politică*, 2014, 5(2):6-7.

Solitude Project, 'Prea-fericitii din groapa de gunoi' [The all-blessed of the landfill], 56', 2011. Available at: https://vimeo.com/32605959.

Şerban, Alina (ed.). *Evicting the Ghost. Architectures of Survival*, Bucharest: Center for Visual Introspection, 2010.

ŞPAC-Şcoala Populară de Artă Contemporană/Protokoll. 'Eşti în Pata Rât' [You are in Pata Rât], in Adrian Dohotaru, Hajnalka Harbula and Enikő Vincze (eds) *Pata*, Cluj: Efes, 2016, pp. 138–162.

Vincze, Enikő. 'The Ideology Of Economic Liberalism And The Politics Of Housing In Romania', *Studia Ubb. Europaea* LXII (2017): 29–54.

Vincze, Enikő and Zamfir, George. 'Racialized housing unevenness in Cluj-Napoca under capitalist redevelopment', *City Journal* 23.4-5 (2019): 439–460.

Voinea, Raluca. 'Public Space' in Tranzit Czechia, *Atlas of Transformation,* 2011. Available at: http://monumenttotransformation.org/atlas-of-transformation/html/p/public-space/public-space-raluca-voinea.html.

Zamfir, George. 'Cluj Capitală Culturală sau justiţie culturală ca soluţie la injustiţii sociale?', *Gazeta de Artă Politică* 15 (2016):12–13.

ART IN THE INTERIM: HOW THE ISSUE OF THE RESTITUTION OF HOUSING IN REUNIFIED BERLIN LED TO AN ARTISTIC REIMAGINING OF THE CITY

NICOLA GUY

...the Germans once again need a bit of glitter in their hovels. [1]

– Franz Hessel

Berlin's relationship to art and creativity is long-standing, with its position as a cultural city arguably being secured in the Weimar period when it became a space of experimentation through the arts. Throughout history, the perception of Berlin as an artistic city has been pushed and used as both a disruptor and an agent of conformity, though it is only since reunification that Berlin has become one of the main centers of the western art world, home to over 400 galleries, the Berlin Biennale, Gallery Weekend and myriad other events and occasions that promote the contemporary art market. [2] Conversations surrounding art's complicity with the gentrification of Berlin are unavoidable, with good reason, as we see rents rise and more and more people being displaced from their homes.

Thinking back from the current perception of the city to look at the immediate period after the reunification of Berlin, we can look at how the relationship between art and urban space was used as a strategy by which the city might reimagine itself. This moment was marked by both uncertainty and change as different individuals and groups attempted to make claims on the city, with the dilapidated housing stock of the central neighborhoods being particularly contested through the controversial process of restitution. Looking at examples of autonomous and institutionally organized exhibitions, the contribution to the changing face of Berlin will be examined, we can seek to understand how these used the period of restitution as a means for their own intentions and the consequences of these efforts.

The dissolution of the German Democratic Republic (GDR) happened quicker and more peacefully than had been anticipated with the fall of the Berlin Wall in November 1989 and the official reunification of the two Germany's just eleven months later in October 1990. The period that followed was chaotic, uncertain and full of a unique kind of energy and excitement that was associated with the rebuilding of the country, with Berlin once again at its center. For some, this excitement was due to the myriad speculative possibilities for redevelopment, regeneration and the opportunity to purchase real estate at cut prices in central locations [3] and for others the reunification was a chance to explore Berlin again and push for a different kind of city.

---

1   Franz Hessel, *Walking in Berlin* (repr., Scribe Publications, 2017), p. 20.
2   Jennifer Allen, 'Made in Berlin', *Frieze D/E*, 2010, https: //frieze.com/article/made-berlin?language=de.
3   Elizabeth Strom and Margit Mayer, 'The New Berlin', *German Politics And Society* 16, no. 4 (1998): 122.

The excitement and push for change were particularly apparent in the central and formerly eastern neighborhoods of Berlin where the majority of housing stock had been state-owned by the GDR. As that state no longer existed, the question of restitution was raised with urgency with the conclusion that the properties either needed to be returned to their original owners or sold off in an attempt to recover some of the debt that the GDR had accrued. Whilst this bureaucratic chaos ensued, in order to solve the situation other additional bids for spaces were made that looked at alternative ways that the new city could be imagined. Artistic activities, including performances and exhibitions, became a way in which these ideas could be highlighted and could utilize the uncertainties of this period in order to reimagine the *Wende* that was quickly becoming a westernization.[4]

Reunification meant that areas which had been on the peripheries of the city found themselves again to be inner city locations. For example, Prenzlauer Berg which had bordered the Berlin Wall on the eastern side was transformed into a central neighborhood again, just north of Mitte, which was also being touted as the centre of Berlin. These areas quickly became the most desirable since the ownership of properties was disputed, markets followed. The changes that would happen in these areas would showcase the new Berlin, be symbolic of reunification and the end of the division, and be scrutinized by the rest of the world.[5] In these circumstances, central Berlin was hotly contested by developers attempting to purchase prime real estate at cheap prices via the controversial Treuhand Agency that administered the privatization of former state owned buildings.[6] As well as others wanting to make a claim to these neighborhoods there were also protests against the redevelopment, with fears it would cause displacement and gentrification leading to Prenzlauer Berg becoming something of a battleground for these debates.[7]

These protests coincided and informed a resurgence in squatting in the city, with individuals and groups taking over the many empty residential and industrial buildings that were particularly plentiful in these newly central neighborhoods. Squatting took on a vital role in housing activism and claimed a key position in the politics of the city as it transformed by causing extensive disruption for the authorities with squats often becoming the sites of contestation or organizing hubs of these protests. Similarly, at this time, there was an increasing number of artistic events being organized alongside restitution, rather than in connection to the regeneration. These activities can loosely be split into either activities that were autonomous or activities that were institutional. Nonetheless, these two different approaches were reunified by their location of empty housing buildings and influence in the landscape of central Berlin neighborhoods. However, on some occasions intentionally, these events intersected with the

4   George J.A. Murray, 'City Building and the Rhetoric of "Readability": Architectural Debates in the New Berlin', *City & Community* 7, no. 1 (2008): 3.

5   Hilary Silver, 'Social Integration In the &Quot; New&Quot; Berlin', *German Politics and Society* 24, no. 4 (2006): 3.

6   Andrej Holm, 'Urban Renewal and the end of Social Housing: The Roll Out of Neoliberalism In East Berlin's Prenzlauer Berg', *Social Justice* 33, no. 3 (2006): 116.

7   Matthias Bernt and Andrej Holm, 'Is it, or is Not? The Conceptualisation of Gentrification and Displacement and its Political Implications in the Case of Berlin Prenzlauer Berg', *City* 13, no. 2-3 (2009): 315.

issues surrounding property ownership and restitution in the former East. Each approach built on Berlin's history as a cultural city and highlighted different possibilities for the neighborhoods that meant that even though temporary they impacted how the city was redeveloped, though with different methods and intentions often in opposition.

Restitution in Contested Times

When the Wall fell, the former east of Berlin was in a state of disrepair, with World War II bomb damage visible across the city and few renovations having been carried out on the older and predominantly residential buildings, of which there were many in the new central neighborhoods. In the further east of the city, Karl Marx Allee, for example, new housing blocks, *Plattenbaus,* had been built that visually altered the landscape. However, in Prenzlauer Berg there had been little updates to the residential buildings during the GDR, the majority of which were *Altbaus*, meaning that despite its dilapidation, it was visually continuous with the western sides of the city. In addition to this, it is estimated in Prenzlauer Berg that 10-20% of residential buildings were empty.[8] The culmination of these conditions and its central location meant that it quickly popularized. It soon became one of the first subjects of a major redevelopment[9] and by 1993 being designated a *Sanierungsgebiete* [urban renewal area].[10] Additionally, when restitution was addressed in Prenzlauer Berg there were few claims made to the buildings, only 35,000 claims being made on residential buildings, out of which 10% were successful.[11] The result of this was that the properties could be, thus allowing the developers who had been keenly watching the progress to step in and make purchases and beginning the process of privatization.

Many Berlin inhabitants became cynical of the *Wende* as it unfolded into westernization rather than a true reunification, which led to increased activism and protest increased across the city and calls for a Third Way being made by a disenfranchised public. This dissatisfaction with the *Wende* was one of the contributing factors that led to a resurgence in squatting in the city, that saw many individuals and groups taking over the empty buildings in the former East Berlin. In the newly squatted buildings, this third way was experimented with as a way of living that drew on ideas from socialism without the authoritarian structure for which the GDR had been notorious.[12] This move by squatters to east Berlin went against the trend of people leaving East Germany for the West in attempts to secure better living conditions,[13] and the non-conformity of moving instead eastwards heightened the anti-establishment atmosphere of the movement.

---

8   Andrej Holm, 'Urban Renewal and the end of Social Housing: The Roll out of Neoliberalism in East
      Berlin's Prenzlauer Berg', *Social Justice* 33, no. 3 (2006): 115.
9   Uta Papen. 'Commercial Discourses, Gentrification and Citizens' Protest: The Linguistic Landscape Of
      Prenzlauer Berg, Berlin'. *Journal of Sociolinguistics* 16, no. 1 (2012): 57.
10   Matthias Bernt and Andrej Holm, 'Is it, or is not?': 316.
11   Mark Blacksell and Karl Martin Born. 'Private Property Restitution: The Geographical Consequences
      of Official Government Policies in Central and Eastern Europe'. *The Geographical Journal* 168, no. 2
      (2002): 185.
12   John Feffer, 'Squat Paradise: East Berlin in the 90s', *Slow Travel Berlin*, 2014, http: //www.
      slowtravelberlin.com/squats-neo-nazis-friedrichshain-in-the-90s/.
13   Esther Peperkamp et al., 'Eastern Germany 20 Years After', *Eurostudia* 5, no. 2 (2009).

Squatting was by no means new to Berlin in the 1990s, the city has a rich history of looking alternatively at housing, and squatting had been present in both east and west during the division. The two sides of the city had different approaches to squatting, and generally, different terms are used to describe each side's approach; *Schwarzwohnen* [illegal living] for the east and *Besetzen* [occupation] for the west.[14] The squatters in the east side had been using squatting as a means of highlighting the poor housing conditions in the eastern properties, that went against the claims of housing being a fundamental right in the GDR. And in western Berlin, which in the 1980s had become an international hub for subcultures, there had been a strong wave of squatting in Kreuzberg, a neighborhood bordering the Wall, during the 1980s.[15] After the fall of the Wall the potential of these scenes to meet and work together in the myriad empty buildings in the east could be explored and, more and more squats, were established across Berlin with Prenzlauer Berg and Friedrichshain being particularly popular.

While these new squats had been relatively left alone between November 1989 and October 1990, when many of the buildings and city institutions were stuck in bureaucratic confusion, the issue of squatting became a focus of the authorities soon after. On the 12th November 1990, a little over a month after Berlin was officially reunified, police visited a row of eleven squatted houses on Mainzer Straße in Friedrichshain, former east Berlin, with the intention of evicting the squatters who had been occupying the houses since the fall of the Wall. The squats on Mainzer Straße occupied a row of houses, around two hundred meters long, all occupied by different groups. Their presence was made known through signage and decoration to the façades, which covered all houses. The squatting scene in Berlin mobilized against the evictions and hundreds took to the street to protest in support of the squats and attempted to save them. For three days, a battle took place on the Mainzer Straße, with around 3000 police officers attempting to raid and clear the squats using violent methods, including employing water cannons and stun grenades.[16] Eventually, the police succeeded in clearing the squats but the events saw over 400 activists being arrested and many casualties on both the side of the squatters and police. There were additionally significant and long term ramifications for the squatting movement more broadly, as both the authorities and activists recognized a need for a change of approach in how they were each working in their own fight for the city.[17]

For the state, the events had been high profile and messy, with previous non-supporters of the squatters making public statements against the police. In a newspaper report made the evening after the demonstrations, a participant is quoted as saying, 'I am not a supporter and I am not a sympathizer of the squatters [...] but I am against violence regardless of the reason'.[18] While these opinions would not have been a concern for the squatters, it was a challenge for the state in this time of political instability. The protests that had led to the fall of the Wall had

14   Alexander Vasudevan, 'Schwarzwohnen: The Spatial Politics of Squatting in East Berlin', *Opendemocracy*, 2013, https://www.opendemocracy.net/en/opensecurity/schwarzwohnen-spatial-politics-of-squatting-in-east-berlin/.
15   Alan Moore, *Occupation Culture*, New York: Minor Compositions, 2015, p. 138.
16   Susan Arndt, *Berlin Mainzer Strasse*, Berlin: BasisDruck, 1992, pp. 10-29.
17   Alexander Vasudevan, *The Autonomus City*, London: Verso, 2017, p. 168.
18   J Tagliabue, 'EVOLUTION IN EUROPE; Berlin is Rocked by a Squatters War', *Nytimes.com*, 1990, https://www.nytimes.com/1990/11/15/world/evolution-in-europe-berlin-is-rocked-by-a-squatters-war.html.

been relatively peaceful from the side of the protesters but violent from the side of the GDR police. This violence had served to gain the Federal Deutsche Republic more support as people wished to move away from the highly policed state.[19] But the return to violence around the Mainzer Straße evictions so soon after reunification worked against this and caused an increasing feeling of public dissatisfaction with the changes being made in the city. This was not helpful for the state, which were attempting to keep reunification as peaceful and smooth as possible in spite of the precarious nature of the transition. Furthermore, the confrontations had caused a split between Berlin's Social Democrats and the Green Party, who had up until this moment been in the coalition, creating fundamental instability within the ruling parties.[20] Not wanting a repeat of Mainzer Straße and wanting to keep public opinion on the side, the authorities needed to work out a different way to deal with what they considered to be the squatting problem.[21]

One of the results of this was that authorities gave some squats the chance to legalize in certain situations if permitted by the buildings' owners. Often squats were able to negotiate a nominal or very low rent in order to remain in the building, though of course this came with conditions and many squatters found themselves being required to carry out substantial renovation work to the buildings in order to remain.[22] As well as this, these contracts were precarious and often short term.[23] The benefits these arrangements had for the authorities are clear; it meant that empty buildings did not remain so, meaning they would be looked after and not at risk from illegal squatting. It also had the benefit of appearing to be an act of generosity, which helped to win back some of the public support lost during the Mainzer Straße protests. However, a harsh view was still taken of illegal squats and legislation was formalized in an attempt to substantially lessen the number in the city. Whilst appeasing the authorities was not the aim or concern of the squatters, remaining in their homes was. Therefore, working within the new legislation was imperative, and though precarious, this was a really viable opportunity for this to happen for the time being.

The Rise of the Art Squat

John Feffer, a former squatter from Mainzer Straße said of the squats, 'Mainzer Straße was also a place of culture and creativity. It was the only colorful street in the whole district. Today Friedrichshain is said to be the creative district. But in 1990 the creative potential was evicted'.[24] This quote highlights the focus on creativity in these squats which were offering an alternative to the dominant discourse of privatization that supported big businesses above the individual. An important factor is how the squats acknowledged their differences to passers-by

19   Wolfgang Mueller, Michael Gehler and Arnold Suppan, *The Revolutions of 1989*, Vienna: Verlag der Österreichischen Akademie der Wissenschaften, 2016, p. 122.
20   Andrej Hold and Armin Kuhn, 'Squatting and Urban Renewal: The Interaction of Squatter Movements and Strategies of Urban Restructuring in Berlin', *International Journal of Urban and Regional Research* 35, no. 3 (2010): 644-658.
21   Alex Vasudevan, *Metropolitan Preoccupations*, Oxford: Wiley, 2015, p. 169.
22   Vasudevan, *Metropolitan Preoccupations*.
23   Sandler, Daniela. *Counterpreservation*. Ithaca: Cornell University Press, 2016, p. 2016.
24   Geronimo, George N Katsiaficas, and Gabriel Kuhn. *Fire and Flames*. Oakland: PM Press, 2012.

by covering the buildings' facades with images, messages in a refusal to blend in with the rest of the houses on the street. This was a marked development from the culture of squatting in East Berlin, where the squatters had previously attempted to go unnoticed, and soon, this visual display of the squats became the norm. This increased visibility made the presence of squats obvious to both anyone passing by in the neighborhoods and to the authorities. This tactic has arguably used a way of staking a claim on the space.

Similar creativity was extended to the activities and organization of the squats themselves, where including cultural events or artistic endeavors was becoming increasingly common. For example, Mainzer Straße was the location of the Tuntenhaus [House of Drag], a queer squat that had already been established in West Berlin in 1981 and later on Kastanienallee, Prenzlauer Berg. The Tuntenhaus organized infamous shows in their backyard and an annual Tuten Festival with drag shows and musical performances.[25] Even better known is the Kunsthaus Tacheles, a ruined building in Mitte that was squatted in 1989 as an art space which quickly became one of the main locations for non-mainstream arts events in Berlin.[26] Tacheles held exhibitions, events and had a cinema, cafe and bar as well as workshops that could be used. It also, importantly, soon established itself as a *Verein* (e.v.) [Association] which was key to its survival when later threatened with demolition and eviction in 1990. After a short debate, the building was secured with a Preservation Order and funding from the Planning Department to renovate the building.[27] Tacheles was one of the longest running spaces of its kind and was eventually demolished in 2012 with a legacy as an important art space in the city. The vibrancy of these spaces attracted an audience and created a lively scene around them. Their presence grew in neighborhoods as they showed the options beyond the privatization of restitution. Visibility was essential for this and further ideas and alternative practices were explored.

The Case of K77

An important moment in the history of the art squat in Berlin was in June 1992 when a group of artists and performers, working under the name Vereinigte Varben Wawavox [*Vereinigte* from the verb *to unite*] took over 77 Kastanienallee, one of the oldest residential building in Prenzlauer Berg. The entire process was artistic in its presentation, with even the procurement of the building taking the form of a piece of a performance art. On the first day of occupation, the group dressed as medical professionals and paraded through Prenzlauer Berg via Kollowitz Platz, where there were more squats, to the empty Kastanienallee 77. The group entered the building and hung signage on the house that declared a medical emergency.[28] The emergency was, they stated, that the house was dying and in need of a heart transplant through the lack of life running through the house. The transplant was complete once they took over occupation and made it safe once more for habitation. They declared that the building was no longer a

25   Azomozox, 'Gender and Squatting in Germany Since 1968', in *Making Room: Cultural Production In Occupied* Spaces, Berlin: Other Forms, 2016, p. 170.
26   Janet Stewart, 'Das Kunsthaus Tacheles: The Berlin Architecture Debate of the 1990s in Micro-Historical Context', in *Recasting German Identity*, London: Boydell & Brewer, 2002, p. 54.
27   Stewart, 'Das Kunsthaus Tacheles', p. 56.
28   Vereinigte Varben Wawavox, 'Squatting Is Art', Pamplet, Berlin, 1992, Papiertigre.

building and was now a 'social sculpture', using artist Joseph Beuys' term, and their actions were performance art. This additionally riffed off the history of Prenzlauer Berg as a cultural neighborhood, often cited as the place to find bohemia in East Berlin.[29]

A week after their occupation, the group behind it distributed a pamphlet about their actions that proclaimed on the front cover 'Squatting is Art', which explained the performance and their intentions with the house. They wrote, 'We have to explain that it is by no means an occupation, but rather an art campaign and that art is under the protection of the Basic Law' (own translation). This decision to announce the project as art rather than a traditional occupation through squatting was conceptual but also served a practical purpose. By saying that their actions were an artistic performance rather than squatting, the group were able to transcend the law that had come into effect in 1990 that decreed that all squats must be cleared by the police within 24 hours of the building's occupation. Art gave the group freedom to occupy the building and from there negotiate their position in the house. The pamphlet goes on to detail what happened next, which was meetings with the Wohnungsbaugesellschaft Prenzlauer Berg [WiP], the housing management firm dealing with the area. The efforts were successful and an agreement was formalized in 1994 when they were granted a 50 year contract to stay in the building with the squat becoming known as K77.[30]

In K77, we see a significant example of how art was being used as a tool with which to reimagine Berlin with art being a method and also the content. This went beyond using the squat just to show art but considered the building and the social relations it contained as art itself. Yet, their aims were relatively modest with organizers being mainly concerned with making the space habitable. Despite modesty, this contrasted with the state's focus on the houses being sold rather than on them being livable spaces. In these arts squats and the case of K77, we see exhibitions and culture being used as a means to push for a wider change in the landscape of Berlin. Even if these spaces were relatively small they were part of a larger network through which the status quo, in this case, the rapid changes happening to the city, could be challenged and those challenging it to find support. These cultural events also contributed to cultural discourses in the city, where we can see other cases of provocations for possible changes to the city being made through the lens of urban space.

Mitte and Prenzlauer Berg, were becoming a hub of events, especially music and parties, quickly transforming it into a fashionable neighborhood for young people, many of whom were involved in working in the creative industries.[31] In addition to the network of squats, there were more clubs, bars and galleries being set up. There was a focus on maintaining freedom around these spaces, which was, of course, a natural reaction to the reunification, it also led to many different activities and community spaces being organized or established in the same areas causing fast-paced changes.[32] The squats and arts events surrounding them

---

29   Philip Brady and Ian Wallace. *Prenzlauer Berg*. Amsterdam: Rodopi, 1995. and Claire Colomb. *Staging the new Berlin*. Hoboken: Taylor and Francis, 2013, p. 7.

30   Ursula Maria Berzborn and Steffi Weismann, *Kule*, Berlin: Revolver Publishing, 2016, p. 371.

31   Jochen Becker, 'New Mitte/Helle Mitte', *Inventory* 4, no. 2 (2001): 52.

32   Anke Fesel and Chris Keller, Berlin Wonderland: Wild Years Revisited 1990-1996, Berlin: bobsairport [etc.], 2014, p. 39.

were autonomous, meaning that they operated independently and then only worked with the authorities once established if it was necessary for their survival. Squats were also, in many cases operating in line with the ideas the autonomous movement; the Marxist ideology that has been born from the 1968 uprisings and become important in counter-mainstream political activities. The West German Autonomen had been involved in squatting in the 1980s, and as the movement progressed after the reunification, the movement continued working against fascism, racism and anti-Semitism, using squats as hubs for these political discussions.[33] These political underpinnings are important when considering practically how these squats operated, with their ideas being in essence against the state and therefore working as an independent.

37 Räume

Squatting not however the only way of accessing space at this time, and there were art events that utilized the empty spaces but procured them through legal means. One key example of this is the exhibition *37 Räume* that was also organized in the summer of 1992 and nearby to Kastanienallee on Augustraße, Mitte. As the title suggests, *37 Räume* was held across 37 different venues along the street with each venue, or room, curated by a different Berlin-based curator, though its concept took its cues from the western art world rather than culturally specific histories of Berlin. There was no overarching theme for the exhibition the intention was to showcase what was happening in the contemporary art scene in Berlin at that moment. The project was conceived by Klaus Biesenbach, founder of Kunst-Werke, the Berlin Biennale and currently Director of the Museum of Contemporary Art, Los Angeles and organized together with Brigitte Sonnenschein. One of the central venues of *37 Räume* was 69 Augustraße, now the home of the internationally known art centre Kunst-Werke (hereafter KW). Similarly to K77 the building of KW, a former, and at that time abandoned, margarine factory, had been empty Biesenbach was approached cultural administration with the suggestion he may take it over and in a similar state of disrepair.[34]

Like many others, Biesenbach and his co-organizers took this moment of chaos as an opportunity to rethink what Berlin could be and use the empty spaces as venues for their suggestions. The fundamental difference with *37 Räume* is that it had been organized with support from the state, unlike any of the squatted art spaces. As well as the installations, there was also an events program that ran alongside the exhibition and this included a curator vs. artists football match. Among the spectators for this was Walter Momper, at the time the Mayor of Berlin, an SPD party member and a figure who had been heavily criticized for his links with the business sector and in the development of east Berlin, a figure, we can assume, would not have been supportive to or a participant in art events held at squats.[35]

33   Geronimo, George N Katsiaficas, and Gabriel Kuhn. *Fire and Flames*. Oakland: PM Press, 2012, p.79.
34   Klaus Biesenbach, '"We Had To Create Something New": Klaus Biesenbach On Inventing The Berlin Biennale', *Artnews.com*, 2018, https://www.artnews.com/art-news/news/create-something-new-klaus-biesenbach-inventing-berlin-biennale-060717-10450/.
35   S Stuk, 'Heftige Kritik an Seinen Kontakten zu Spreepark-Interessenten: Mompers Geschäfte Passen auch den Genossen Nicht', *Berliner Zeitung*, 2013, https://www.berliner-zeitung.de/heftige-kritik-an-seinen-kontakten-zu-spreepark-interessenten-mompers-geschaefte-passen-auch-den-genossen-

In his introduction to the project in its catalogue Biesenbach writes that the use of the rooms being used was currently in question, suggesting that there was uncertainty in the future of the empty spaces, and that their use was up for debate.[36] Arguably this is similar to the motives of the organizers of K77, who were also using the uncertain state the building was in as a lever to call for its reevaluation. However, there was a difference in what they were suggesting and this played out through their activities. Despite using arts activities, the organizers of K77 were not suggesting that the buildings should be used exclusively as venues for art. The act of living in the space was still the main aim. In marked contrast, in his approach to *37 Räume*, Biesenbach was clearly raising the question of whether the predominantly residential buildings in which the exhibition took place should remain residential at all.

The exhibition *37 Räume* was realized with the support of Jutta Weitz who worked for the housing authority and used her position to support arts initiatives in the areas around Mitte by allowing temporary use of empty spaces. Likened to Joan of Arc and called the 'key figure behind Mitte's cultural development',[37] Weitz indicates the central presence of individuals within the authorities who wished to support something other than privatization. Indeed, Biesenbach cites Weitz as persuading him to use 69 Augustraße so the building could avoid becoming another gym.[38] However, we must also look at what the consequences of projects such as these were and *37 Räume* has been attributed to being the catalyst for the development of Augustaße, now entirely gentrified and one of the main streets for commercial galleries in the city. The spaces were only given over temporarily meaning that the exhibition was only open for a week. This was a relatively short time for it to be able to prove its value though it proved to be popular with some 35,000 people [39]visiting the exhibition and attending the opening, which was compared to a festival taking over the street.[40] The opening had been planned to coincide with the ninth iteration of the quinquennial exhibition Documenta, which takes place in Kassel, with the hopeful plan being that the audience would travel over to Berlin after the main event which worked and the exhibition and its organizers gained international attention. Like the Mainzer Straße protests had been a pivotal moment in the rethinking of the Berlin squatting scene, *37 Räume* had a similar effect on the Berlin contemporary art scene and remains an oft-cited reference in how Berlin was transformed into being a major player in the art world.[41]

nicht-15781990.

36   Klaus Biesenbach, *37 Räume*, Berlin, 1992, p. 7.

37   Anke Fesel and Chris Keller, *Berlin Wonderland: Wild Years Revisited 1990-1996*, Berlin: bobsairport [etc.], 2014: 178.

38   Klaus Biesenbach, 'Klaus Biesenbach Recalls the Founding of KW in Berlin 25 Years Ago, a Moment of "Radical Change and Freedom"', *Artnews.com*, 2016, https://www.artnews.com/art-news/news/klaus-biesenbach-recalls-the-founding-of-kw-in-berlin-25-years-ago-at-a-moment-of-radical-change-and-freedom-7370/.

39   Klaus Biesenbach, 'Klaus Biesenbach Recalls the Founding of KW in Berlin 25 Years Ago, a Moment of "Radical Change and Freedom"'.

40   Anke Fesel and Chris Keller, *Berlin Wonderland: Wild Years Revisited 1990–1996*, Berlin: bobsairport [etc.], 2014: 71.

41   Tara Mulholland, 'Berlin: Once East German Gritty, Now Slick, But Still Artsy', *Nytimes.com*, 2010, https://www.nytimes.com/2010/12/18/arts/18iht-scberlin18.html.

Despite these successes, *37 Räume* was also subject to a number of protests throughout the short week that it occupied Augustaße. Its presence on the street – the signage and many visitors – were interrupted through interventions by groups that disagreed with the exhibition and what it was perceived to stand for. Photographs taken at the exhibition,[42] show posters were stuck over the branding with slogans such as '37 without rooms' [37 Ohne Räume], 'Art Whore' [Kunst Nutte] and labels with 'Room 38' [Räume 38] were stuck to trash cans. The inhabitants of a building not involved with the exhibition affixed a sign reading 'No art here today. 3 Easterners',[43] which mirrors Biesenbach's reflections that 'all the area residents—and the organizers—were very happy that it wouldn't continue'.[44] Reading these slogans, we can understand them as countering the intentions of the exhibition and considering this in light of the housing crisis that many were facing at this moment, and these statements asked the question; why fill empty houses with art when there are so many without spaces to live in?

The aim of the *37 Räume* was to showcase the contemporary art scene of Berlin, not just to the people living in Berlin and involved with the scene but with the wider aim of bringing a new audience to the city; it was thus for a different community as well. This differed from the work of the squats, who were looking to provide support for those already in the city and reimagine how it could be for them rather than how space could be used to benefit a new audience.

As indicated above *37 Räume* has had a clear and, in many ways, successful legacy. It paved the way for the Berlin Biennale, which was set up by Biesenbach and other curators in 1996 and held its tenth iteration this summer in 2018. KW is an internationally renowned arts institution that shows a number of high-profile exhibitions each year, often to critical acclaim. Augustaße itself has been transformed into a desirable street full of commercial galleries, restaurants and expensive clothing stores. Both KW and the Berlin Biennial, which until this iteration was organized by KW though they are now separate entities, receive regular funding from the Bundestag as well as a host of other sources, including BMW. It is also demonstrated that contemporary art could bring a new audience to the city and their money with it.

Conclusion

Later, when the Treuhand Agency had closed, and it was considered this part of reunification was finished, the image of the city went through an overhaul as the 'new Berlin' was present-ed through advertising and strategies of place marketing in order to reestablish tourism to the city.[45] In this reimagining, art and culture were key, and Berlin was presented as a go-to destination for this. After reunification, the image of Berlin went through an overhaul as the 'new Berlin' was presented time and again through place marketing in order to reestablish tourism in the city. This included the utilization of DIY and artistic aesthetics and the politics that went alongside that but without attempting to engage with those politics, often cynically

---

42   I am indebted to Klaus Baedicker for sharing with me the photographs he took of *37 Räume*.
43   Fesel and Keller, *Berlin Wonderland: Wild Years Revisited 1990-1996*, p. 181.
44   Biesenbach, 'Klaus Biesenbach Recalls the Founding of KW in Berlin 25 Years Ago, a Moment of "Radical Change and Freedom"'.
45   Claire Colomb, *Staging the new Berlin*, Hoboken: Taylor and Francis, 2013, p. 26.

perceived by those who had been involved with the squats or art spaces whose images were being reproduced.[46] The activities in the early years of reunification that had utilized art as a tool for reimagining the city had clearly had an impact within wider Berlin. Whilst in the early 1990s it was possible to act autonomously to acquire space and set up venues that challenged in some way the status quo this has clearly been less possible as time has gone on.

Restitution had resulted in privatization and empty houses became fewer, affecting the potentials for squatting as the crack-down in laws had as well. Even those of the spaces that have managed to remain have had to radically change their organizations and are no longer able to operate. However, what was important about these spaces was that they experimented with an alternative and used creativity to force their visibility into Berlin neighborhoods and the history of this should remain important today. As Alexander Vasudevan writes, '[…] in the case of contemporary Berlin, a stronger awareness of these histories might still point us to an alternative beyond a city increasingly shaped by the logics of profiteering and privatization, displacement and dispossession'.[47] While it is no surprise that it was the commercially oriented ventures that were successful and the autonomous spaces that were closed down, all of these activities utilized the chaos of reunification to present alternatives at a time when it was possible to see them more easily. When comparing these approaches, it is, of course, far too simplistic to suggest that 'activists using art = good' and 'curators using art = bad', or to go further and suggest 'art = gentrification' no matter who is the organizer. This article proposes a critical look at how art has been used within the changes to neighborhoods to Berlin in order to consider how we may use it as a tool in future debates surrounding the city. In the same way that all of these events called upon different histories, we can call upon them today in order to reimagine art and urban life.

References

Allen, Jennifer. 'Made in Berlin'. *Frieze D/E*, 2010. https://frieze.com/article/made-berlin?language=de.

Arndt, Susan. *Berlin Mainzer Strasse*. Berlin: BasisDruck, 1992.

Azomozox. 'Gender And Squatting in Germany Since 1968' in *Making Room: Cultural Production in Occupied Spaces*. Other Forms, 2016.

Becker, Jochen. 'New Mitte/Helle Mitte'. *Inventory* 4, no. 2 (2001): 48 -66.

'Berlin Besetzt'. Berlin-Besetzt.de, 2015. http://www.berlin-besetzt.de.

Bernt, Matthias, and Andrej Holm. 'Is it, or is Not? The Conceptualisation Of Gentrification and Displacement and its Political Implications in the Case of Berlin-Prenzlauer Berg'. *City* 13, no. 2-3 (2009): 312–324.

Berzborn, Ursula Maria, and Steffi Weismann. *Kule*. Berlin: Revolver Publishing, 2016.

Biesenbach, Klaus. *37 Räume*. Berlin, 1992.

---

46   Colomb, *Staging the new Berlin*, p. 240.
47   Alexander Vasudevan, 'Schwarzwohnen: The Spatial Politics of Squatting in East Berlin', *Opendemocracy*, 2013, https://www.opendemocracy.net/en/opensecurity/schwarzwohnen-spatial-politics-of-squatting-in-east-berlin/.

Biesenbach, Klaus. '"We had to Create Something new": Klaus Biesenbach on Inventing The Berlin Biennale'. *Artnews.com*, 2018. https: //www.artnews.com/art-news/news/create-something-new-klaus-biesenbach-inventing-berlin-biennale-060717-10450/.

Biesenbach, Klaus. 'Klaus Biesenbach Recalls the Founding of KW in Berlin 25 Years ago, A Moment of '"Radical Change and Freedom"'. *Artnews.com*, 2016. https: //www.artnews.com/art-news/news/klaus-biesenbach-recalls-the-founding-of-kw-in-berlin-25-years-ago-at-a-moment-of-radical-change-and-freedom-7370/.

Blacksell, Mark, and Karl Martin Born. 'Private Property Restitution: The Geographical Consequences of Official Government Policies in Central and Eastern Europe'. *The Geographical Journal* 168, no. 2 (2002): 178-190.

Brady, Philip, and Ian Wallace. *Prenzlauer Berg*. Amsterdam: Rodopi, 1995.

Colomb, Claire. *Staging the New Berlin*. Hoboken: Taylor and Francis, 2013.

Feffer, John. 'Squat Paradise: East Berlin in the 90s'. *Slow Travel Berlin*, 2014. http: //www.slowtravelberlin.com/squats-neo-nazis-friedrichshain-in-the-90s/.

Fesel, Anke, and Chris Keller. *Berlin Wonderland: Wild Years Revisited 1990-1996*. Berlin: bobsairport [etc.], 2014.

Geronimo, George N Katsiaficas, and Gabriel Kuhn. *Fire and Flames*. Oakland: PM Press, 2012.

Hessel, Franz. *Walking in Berlin*. London: Scribe Publications, 2017.

Hold, Andrej, and Armin Kuhn. 'Squatting And Urban Renewal: The Interaction Of Squatter Movements And Strategies Of Urban Restructuring In Berlin'. *International Journal of Urban and Regional Research* 35, no. 3 (2010): 644-658.

Holm, Andrej. 'Urban Renewal And The End Of Social Housing: The Roll Out Of Neoliberalism In East Berlin's Prenzlauer Berg'. *Social Justice* 33, no. 3 (2006): 114-128.

Hoogenhuijze, Leendert van, Bart Van der Steen, ask Katzeff, and Leendert Van Hoogenhuijze. *The City Is Ours*. Oakland, CA: PM Press, 2014.

KuLe. 'Facade - Kunsthaus Kule'. *Kunsthauskule.De*, 2020. http: //kunsthauskule.de/Facade.

KW. 'About – KW Institute for Contemporary Art'. KW Institute for Contemporary Art, 2020. https: //www.kw-berlin.de/en/about/.

Moore, Alan. *Occupation Culture*. New York: Minor Compositions, 2015.

Mueller, Wolfgang, Michael Gehler, and Arnold Suppan. *The Revolutions Of 1989*.

Vienna: Verlag der Österreichischen Akademie der Wissenschaften, 2016.

Mulholland, Tara. 'Berlin: Once East German Gritty, Now Slick, But Still Artsy'. *Nytimes.com*, 2010. https: //www.nytimes.com/2010/12/18/arts/18iht-scberlin18.html.

Murray, George J.A. 'City Building and the Rhetoric of "Readability': Architectural Debates" in the New Berlin'. *City & Community* 7, no. 1 (2008): 3-21.

Papen, Uta. 'Commercial Discourses, Gentrification and Citizens' Protest: The Linguistic Landscape Of Prenzlauer Berg, Berlin'. *Journal of Sociolinguistics* 16, no. 1 (2012): 56-80.

Peperkamp, Esther, Malgorzata Rajtar, Irene Becci, and Birgit Huber. 'Eastern Germany 20 Years After'. *Eurostudia* 5, no. 2 (2009).

Rost, Andreas, Annette Gries, and Heinz Havemeister. *Tacheles - Alltag Im Chaos*. Berlin: Elefanten Press, 1992.

Sandler, Daniela. *Counterpreservation*. Ithaca: Cornell University Press, 2016.

Schlaeger, Hilke, and Nancy Vedder-Shults. 'The West German Women's Movement'. *New German Critique*, no. 13 (1978): 59.

Silver, Hilary. 'Social Integration in the &Quot;New&Quot; Berlin'. *German Politics and Society* 24, no. 4 (2006): 1-48.

Stewart, Janet. 'Das Kunsthaus Tacheles: The Berlin Architecture Debate of The 1990S In Micro-Historical Context'. In *Recasting German Identity*. London: Boydell & Brewer, 2002.

Strom, Elizabeth, and Margit Mayer. 'The New Berlin'. *German Politics and Society* 16, no. 4 (1998): 122-139.

Stuk, S. 'Heftige Kritik an Seinen Kontakten zu Spreepark-Interessenten: Mompers Geschäfte Passen auch den Genossen Nicht'. *Berliner Zeitung*, 2013. https: //www.berliner-zeitung.de/heftige-kritik-an-seinen-kontakten-zu-spreepark-interessenten-mompers-geschaefte-passen-auch-den-genossen-nicht-15781990.

Tagliabue, J. 'EVOLUTION IN EUROPE; Berlin is Rocked by a Squatters War'. *Nytimes.com*, 1990. https: //www.nytimes.com/1990/11/15/world/evolution-in-europe-berlin-is-rocked-by-a-squatters-war.html.

Vasudevan, Alexander. 'Schwarzwohnen: The Spatial Politics of Squatting in East Berlin'. Opendemocracy, 2013. https: //www.opendemocracy.net/en/opensecurity/schwarzwohnen-spatial-politics-of-squatting-in-east-berlin/.

Vasudevan, Alexander. *The Autonomus City*. London: Verso, 2017.

Vasudevan, Alex. *Metropolitan Preoccupations*. Oxford: Wiley, 2015.

Vereinigte Varben Wawavox. 'Squatting Is Art'. Pamplet. Berlin, 1992. Papiertigre.

THE EXPLOITATION OF ISOLATION: URBAN DEVELOPMENT AND THE ARTIST'S STUDIO

JOSEPHINE BERRY & ANTHONY ILES

Have you heard about this house

Inside, a thousand voices talk

And that talk echoes around and around

The windows reverberate

The walls have ears

A thousand saxophone voices talk

You should hear how we syllogize

You should hear

About how Babel fell and still echoes away,

How we idolize,

Theorize

Syllogize

In the dark,

In the heart

— Pere Ubu, 'Dub Housing'

In 2010, in the wake of the 2007 subprime and 2008 financial crisis we wrote, No Room to Move: Radical Art in the Regenerate City, a book which assessed the growing role of public art in urban design in the United Kingdom post-1945 and involved discussions with contemporary artists critical of urban regeneration processes.[1] We anticipated that

---

1 Josephine Berry Slater and Anthony Iles, *No Room to Move: Radical Art and the Regenerate City*, , London: Mute Books, 2010.

the cultural benefits promised by regeneration schemes would be progressively dumped in the era of austerity, as the developers' naked profit principle became an acceptable and open objective, positioning artists increasingly as collateral rather than agents of urban change, and art became a vernacular veneer to be cloned by developers and local authority bureaucrats alike.

However, the reignition of the housing market triggered by the subprime crisis blew away even our worst expectations. The wholesale commodification of urban space in the, by now very much global, city of London, which rapidly escalated the economic crisis into the current housing crisis, has impacted artists' living conditions, working conditions, art practice and the public display of art more generally.[2] Yet how is the spatial precarity that has resulted from this unprecedented transformation of real estate into the main lever of the British economy, and worsened by austerity, made legible in the field of contemporary art? In other words, how does a scarcity of space or spatial scarcity – which, as a fundamental use value necessarily affects all of social production and reproduction – become a directly legible influence on art, both historically and today, and with what effects? The intensifying struggle over housing and workspace must surely manifest in art's internal development as much as in the more externally legible forms of social contestation and organization that involve – but as often implicate – artists. Here, therefore, we attempt to construct a brief overview of the relationship between the urban mode of production, the studio, the social figure of the artist and the nature of their practice.

Here we continue to develop a framework by which we understand art as developing both in relation to and distinction from capitalism's spatial fixes.[3] In David Harvey's analysis, capitalism both fixes space for value production, and then later disaggregates it in order to provide for new areas of innovation, opportunity and profit.[4] Capital's needs for transportation, communication and storage structure space and the environment.[5] In the post-war period, particularly under the pressures of reconstruction, this restructuring was undertaken by the state as a unified programme of public works, within which art was integrated, for the first time, as exceptional and autonomous, serving the purpose of no purpose e.g. spiritually edifying public art, albeit within a context tailored to the

---

2   Of course, there has been a *housing crisis* in London and the UK for most of the twentieth century, arguably for the entire course of modernity, and this is directly linked to the capitalisation of land in the United Kingdom which laid the basis for, and is tied up with, London's centrality to the global accumulation of financial, industrial and landed capital. For a series of cogent arguments about the longevity of the British economy's relation to a 'history of residential inflation', questions about the term *crisis* and the necessity of high land and house prices to the health of UK 'state political authority' see Danny Hayward, 'Fire in a Bubble', *Mute*, 15th September 2017, available at: http://www.metamute.org/editorial/articles/fire-bubble [Accessed 20 August, 2018].

3   See: David Harvey's 'The Spatial Fix: Hegel, von Thunen and Marx', *Antipode*, Vol.13, no. 3, pp. 1–12, 1981 and 'Globalization and the 'Spatial Fix', *Geographische Revue*, No.2, 23-30, 2001.

4   Harvey, 'The Spatial Fix: Hegel, von Thunen and Marx', op. cit., p. 25.

5   Costas Lapavitsas, 'Financialisation, or the Search for Profits in the Sphere of Circulation', London: SOAS, 2009, Available at: http://www.soas.ac.uk/rmf/papers/file51263.pdf. [Accessed 20 August, 2018]

needs of industrial capitalist development.[6] Later, capital's need for spatial fixes become the response to crises of overaccumulation and temporary solutions to the destructive effects of competition – driving the 'annihilation of space by time'[7] – achieved by earlier revolutions in transport and communications.[8] Spatial commodification develops from being a corollary of industrial development with some speculative outcomes (railways and real estate in the 19th century) into a core area of accumulation[9] in the late-20th and early 21st centuries due to the increasing prominence of global finance. It is through this dynamic, according to Michael Hudson[10] and Loren Goldner,[11] that finance ceases to serve investment in productive industry and becomes the central motor of (and in fact barrier to) development as financialization determines the form through which first corporations, then almost all enterprises of every form and function, both large and small, reproduce themselves. Within this emergent field of intense global competition between cities, vying to solicit investment from the swarm of nomad dollars[12] seeking profit worldwide, urbanization becomes a key mediator,[13] and art is at stake within it because as Harvey argues, 'claims to uniqueness and authenticity can best be articulated as distinctive and non-replicable cultural claims.'[14] This provides us with the credible linkages between finance and urbanization which help to structure our framework, as well as the periodization of art within capital's self-development.

Here, we prioritize the studio as a spatial container within which the current production conditions of art are crystallized. The studio offers a window onto the life of the artist in all its distinctness from other working practices in the city which makes it highly desirable to processes of commodification. It is also a spatial frame that allows us to track the totality of the artist's activities. Therefore, the studio presents a surface upon which are etched capitalist financialization's desires for exemplary creative practices, embodied in artists, and at the same time, the minimum conditions required by artists to actually create work. By tracking the historical transformations of the artist's studio we can illuminate the impact of the changing mode of urban production on the figure and practices of artists and thereby sharpen our reading of the effects of spatial crisis on art today. Within this we understand art's negotiation of its spatial conditions as a struggle to both reproduce the

---

6   See, for example, the debates around the exclusion of industrial art in the post-war formation of the Arts Council of England (See: Michael T. Saler, *The Avant-Garde in Interwar England: Medieval Modernism and the London Underground*. New York: Oxford University Press, 2001, pp. 167–169). Notable here is John Maynard Keynes dual role in the integration of art into the state as exception to the division of labour in society and his central importance to the economic reform of the post-war State generally.
7   Karl Marx, *Grundrisse: Foundations of the Critique of Political Economy*, Trans. Martin Nicolaus, New York: Vintage, 1973, p. 538.
8   Henri Lefebvre, *The Production of Space*. Trans. D. Nicholson-Smith. Oxford; Cambridge, Mass.: Blackwell, 1991.
9   Lapavitsas, 'Financialisation, or the Search for Profits in the Sphere of Circulation'.
10   Michael Hudson, *Super Imperialism: the Origin and Fundamentals of U.S. World Dominance,* London; Sterling, Va.: Pluto Press, 2003.
11   Lauren Goldner, 'Fictitious Capital for Beginners', *Mute*, Vol.2 No.6, 2007.
12   Goldner, 'Fictitious Capital for Beginners'.
13   Louis Moreno, 'The Urban Process Under Financialised Capitalism', *City*, Vol.18 No.3, 2014, pp. 244–
14   David Harvey, 'The Art of Rent', *Socialist Register*, 2002, pp. 93–102, p. 98.

artist and reproduce art. This is not commensurate with class struggle, but, given art's difference from processes of capitalization – art is neither defined directly by socially necessary labor nor utility,[15] and derives its force through a critical remove from the status quo of any given societal formation[16] – art's spatial struggles overlap both with other struggles for social reproduction and critiques developed by antisystemic movements.

Five Stages in the Studio's Genealogy

In order to set art's defining relationship to finance in a broader historical context and to tease out this and other factors structuring its relationship to space and the city, we have constructed a comparative genealogy of the artist's studio in modernity. As Daniel Buren commented in 1979: 'Analysis of the art system must inevitably be carried on in terms of the studio as the unique space of production and the museum as the unique space of exposition.'[17] If, as a man of his institution-critical times, Buren focused on exposing the studio's continuity with the museum as the artwork's intended destination and implicit limit condition, we will adapt this connective approach to our own times, replacing the museum with the city.

The museum, as a space apart, functioned as a laboratory for concentrating knowledge during the Enlightenment as well as reflecting the rigid divisions of production within the first phase of capitalism. But within post-Fordism, the city becomes the factory (of creative labor and financial self-valorization) and the crucible of knowledge production which, like the brain itself, thrives on its multiplicity of interactions.[18] If we want to read the studio as a vessel connecting the (financialized) transformation in shifting production conditions to changing conceptions of art and its exhibition which accompany this process, it then becomes necessary to formulate a genealogy in keeping with Michel Foucault's epistemological strategy.

Drawing histories is emphatically not, he argued, about creating consoling continuities between the past and the present moment wherein we rediscover ourselves, but rather a way to 'introduce discontinuity into our very being', to uproot our presuppositions. 'This', he concludes, 'is because knowledge is not made for understanding; it is made for cutting.'[19] By returning to the past we want to emphasize that history is not only a knife to cut backwards through time, but also forwards into the virtual futures stored in the present. In presenting the following brief genealogy of historical models of the studio, we want to emphasize not only what has changed, but also the many historical elements

---

15 Immanuel Kant, *Critique of Judgment*, trans. Werner S. Pluhar, Indianapolis: Hackett, 1987 and Dave Beech, *Art and Value: Art's Economic Exceptionalism in Classical, Neoclassical and Marxist Economics*, Leiden: Brill, 2015.

16 Theodor W. Adorno, *Aesthetic Theory*, trans. Robert Hullot-Kentor, London; New York: Continuum Press, 1997.

17 Daniel Buren, 'The Function of the Studio'. *October*, Vol.10, 1979, pp. 51–58, p. 51.

18 Jane Jacobs, *The Death and Life of Great American Cities*, New York, Random House, (1961) 1993.

19 Michel Foucault, 'Nietzsche, Genealogy, History', *The Foucault Reader*, Paul Rabinow ed., London: Penguin Books, 1984, p. 88.

which persist within current studio culture and architecture. While these presiding elements inform the condition of the contemporary studio, they gain a new function and meaning under present conditions, bearing only a resemblance to their former identity.

Fig.1: Dante Gabriel Rossetti and Theodore Watts-Dunton in Rossetti's Studio by Henry Treffry Dunn, 1882, Gouache and watercolour on paper now on card, National Portrait Gallery.[20]

The Isolated Studio (c.1800 – 1950)

The isolated studio arises with the modern metropolis and the solidification of capitalist society with processes of urbanization. It is a cell, withdrawn from but surrounded by the bustle of the city. Often situated at the city's fringes or derelict zones, partially rural or pastoral in character, it is the situation of the existential artist, laboring over their metier in retreat.[21] Yet, though the studio may contain a particle of the pastoral, its eyrie-like remove from the city also provided a

20 License: https://creativecommons.org/licenses/by-nc-nd/3.0/.
21 Buren, 'The Function of the Studio'. Honoré De Balzac immortalised the archetypal inwardness and nobility of the artist's studio in his 1845 short story, 'The Unknown Masterpiece'. Here is the scene where the young painter Nicolas Poussin first encounters Master Porbus's studio: 'All the light in the studio came from a window in the roof, and was concentrated upon an easel, where a canvas stood untouched as yet save for three or four outlines in chalk. The daylight scarcely reached the remoter angles and corners of the vast room; they were as dark as night, but the silver ornamented breastplate of a Reiter's corselet, that hung upon the wall, attracted a stray gleam to its dim abiding-place among the brown shadows. [...] The walls were covered, from floor to ceiling, with countless sketches in charcoal, red chalk, or pen and ink. Amid the litter and confusion of color boxes, overturned stools, flasks of oil, and essences, there was just room to move so as to reach the illuminated circular space where the easel stood. The light from the window in the roof fell full upon Porbus's pale face and on the ivory-tinted forehead of his strange visitor.' (Honoré De Balzac, 'The Unknown Masterpiece', 1845, available at: https://www.gutenberg.org/files/23060/23060-h/23060-h.htm#link2H_4_0002.)

vantage point from which to reflect upon and unveil its mysteries. The studio exacts its charge precisely from its distant proximity to the people and things that bustle and bristle past. From it, art springs out to bring revelations about urban life into appearance. In early post-studio artist Daniel Buren's condensation:

1. It is the place where the work originates.

2. It is generally a private place, an ivory tower perhaps.

3. It is a stationary place where portable objects are produced.

[...] the studio is a place of multiple activities: production, storage, and finally, if all goes well, distribution. It is a kind of commercial depot.[22]

Although modern artists, such as Edouard Manet and Claude Monet, made forays into the city and country, to paint en plein air, they set out from the studio or reassembled it outdoors (Monet had his Bateau Atelier so he could paint the light-industrial river life at Argenteuil), returning to the isolated studio to complete their work. By the mid-19th century, the artist's studio perhaps already began to seem like a colonial outpost or hunting hide from which to launch explorations into the unknown and increasingly far-flung territories generated by the dynamic forces of finance's first urban transformations. Indeed, it was Baron von Haussmann's credit-fueled redevelopment of Paris (1853-1870) that pushed out manufacturing and the working classes from the city center, creating industrial suburbs such as Argenteuil (where Monet lived and worked), Courbevoie, and Asnières-sur-Seine (where Seurat painted factory workers relaxing) whose hybridization of gritty industrial production and pleasure seeking fascinated the Impressionists. As the old artisanal quarters were dismantled, where housing and manufacture had mixed and self-organized along with the classes, they were replaced by the repetitive uniformity of the new Paris – kiosk, bench, street-lamp, kiosk, bench, street-lamp – ubiquitously evoking industrial standardization and efficiency, not to mention a more rigid spatial division of classes. We can trace the consequences of early financialization's spatializing effects in the siting of the Impressionist studio within the new suburban peripheries, the intensification of speed and mobility, and the attention to the drama of class differences in their paintings.

With the old quarters and lifestyles gone, there was a brash proletarianization of the freshly built public sphere producing a caustic shock to bourgeois sensibilities and, it should be added, creating perhaps the most important subject of 19th century painting and writing. The shock is tangible in the following diary entry by the bohemian bourgeois Goncourt brothers of 18 November 1860:

Our Paris, the Paris where we were born, the Paris of the way of life of 1830 to 1848, is passing away. Its passing is not material but moral. Social life is going through a great evolution, which is beginning. I see women, children, households, families in this café. The interior is passing

22   Buren, 'The Function of the Studio', p. 51.

away. Life turns back to become public. The club for those on high, the café for those below, that is what society and the people are come to.[23]

The first speculative housing boom was paid at the cost of the displacement and broken autonomy of the city's working class, but it also unleashed the anomic bacchanal of mass leisure, giving the petit bourgeois new standing in a reborn public realm. While this mass appearance certainly led to the temporally disjointed flâneur and the probing bateau atelier,[24] it caused an equal and opposite reaction, necessitating the romantic retreat into the existential gloom of the studio in which modernity's headlong transformations could be sifted and digested as art. If artists could be existentially gloomy, however, it was because their studios afforded at least some privacy; the very thing that will become unavailable in the neoliberal city's cloning of the artist's atelier and developers' penchant for large glazed facades behind which there is no place to hide and for which there are no curtains big enough! But 19th century Parisian gloom was perhaps more psychological than architectural since it was there, between 1900 and the 1930s, that the maison atelier or studio-house[25] was invented with its top-floor studios and large windows with living quarters below.[26] The atelier model of light, spacious and multipurpose living spaces would be integrated into modernist open-plan principles of design (e.g. Le Corbusier's Unité d'Habitation, 1947-1952), driving a reimagination of lifestyle that has underwritten middle class appropriations of loft living from the 1970s until today.

Fig.2: Fluxhouse Cooperative 1967–89, Fluxhouse by edenpictures is licensed with CC BY 2.0.[27]

23   Edmond and Jules Goncourt quoted in T.J. Clark, *The Painting of Modern Life: Paris in the Art of Manet and his Followers*, Princeton: Princeton University Press, 1999, p. 34.
24   Monet had a bateau atelier or studio boat at Argenteuil in which Édouard Manet famously painted a portrait of him working accompanied by his wife, *Claude Monet peignant dans son atelier* (1874).
25   The modernist studio-house Zukin refers to is Le Corbusier's Maison-atelier du peintre Amédée Ozenfant, Paris, France, 1922. This itself is a significant upscaling of features drawn from 19th century light-industrial artisans' workshops, often located in the suburbs and working class areas of Paris, used as live-work studio spaces by modernist artists such as Alberto Giacometti, Amadeo Modigliani and Pablo Picasso.
26   Sharon Zukin, *Loft Living: Culture and Capital in Urban Change*, Baltimore: Johns Hopkins University Press, 1982, pp. 80-81.
27   To view a copy of this license, visit https://creativecommons.org/licenses/by/2.0/.

The Factory Studio (1962-1989)

Warhol's Factory studio helped pioneer the appropriation of former industrial space by artists in Manhattan as the city stepped up its planned displacement of industry.[28] The studio as factory expresses a dual process of the western city's deindustrialization and art's experimental reimagination of itself as participating in general production: collective, industrial, site-sensitive, gendered, prototyped, banalised, libidinal, democratic and mediatized. The factory studio was a space for collapsing distinctions between art and production, art and life, art and technological reproduction – a laboratory for the contestation of art's distinction per se and the creation of large-scale, dirty, genre shifting, commerce courting, performative and intermedia works. For the generation that followed the Abstract Expressionists, the factory studio became a space not only for the reinvention of art, but for the reinvention of the self, which in turn was framed as art and put on display: 'In the mid- to late sixties, for example, the Conceptualists presented the process of making art as a work of art in itself.'[29] In this phase, the neo-avant-garde renewed art by challenging its originality, individualism, sexism and class elitism. The factory was therefore a space of liberated production in which to launch attacks on the lazy presumptions of an art system that had been embalmed within the museum and bourgeois class interests. The bones of working-class production were danced on, not in the sense that artists directly displaced industrial labor, but rather that their joyful occupation of these decommissioned spaces examined and overturned many of the social divisions that industrial production presupposes.

Artists also organized, as 'Art Workers', on the model of industrial labor but not (primarily) to 'campaign about wages or working conditions',[30] but instead to radically contest the disaster of bourgeois capitalism, racism, war and technocracy.[31] From Pop, to Performance and Land Art, artists were at pains to make explicit the relationships between art and commodification, art and the forces of production (industrial, administrative, cybernetic, heteropatriarchal). Like a can of soup, artworks could be generated by the dozen, sold by the yard, and everyone could be a celebrity for 15 minutes. Equally, the productive power of the factory could be desublimated, its repetitions and intense energies hijacked to produce new experimental gyrations of thought, sex, art, music and performance. Its expanse of largely unarticulated space provided an (almost) blank sheet for rearranging the conventions of living, working and creating (live/work).

---

28   Of course this was not the first foray of artists into the factory, in the 1920s Russian constructivists had developed first a *laboratory productivism* in the studio before launching themselves into *practical work in production*. However, as Maria Gough and others' key studies indicate, the artist primarily entered the factory either as an educator, designer of products or re-trained engineer. See, Maria Gough, *Artist as Producer Russian Constructivism in Revolution*, Berkeley, Calif.: Univ. of California Press, 2007, pp. 100–119.

29   Zukin, *Loft Living*, p. 80.

30   Julia Bryan-Wilson, *Art Workers: Radical Practice in the Vietnam War Era*, Berkeley: University of California Press, 2009, p. 114.

31   Julia Bryan-Wilson's book cogently and sensitively discusses the formation of the Art Workers Coalition and its offshoot, The New York Art Strike against Racism, War and Repression, during the early 1970s in terms which bring the politics of labour and art-labour to the fore in their complex and contradictory web of associations, identifications and disidentifications.

One of the most ambitious artists' housing projects, George Macunius' Fluxhouse Cooperatives (1966 to 1975), crosses over with Warhol's Factory, sharing the conditions of New York's deindustrialization in the late '60s and early '70s, but put them to use under a very different model. Whilst Warhol's Factory has latterly been celebrated as a living artwork and business model, Fluxhouse Cooperatives were conceived as a multiple from the beginning. They were a bohemian capitalist fantasy of a communist cooperative trading under the name Fluxus Cooperative Inc.[32] Maciunas imagined Fluxhouses or the Fluxcity (of which the coops were the basic building blocks to be multiplied and scaled-up) as a Kolkhoz (collective estate). Yet the pragmatism of negotiating New York's zoning laws and the difficulty of raising finance meant this was to become only replicable on an individualist and increasingly capitalized basis by an incoming middle class and the developers that followed hard on their heels.

For a short period during the 1960s and early 1970s then, Loft Living,[33] was a cheap fix for artists and a radical new way to think about creating and living in the city – one that inspired many imitators. Zukin argues, however, that through the growing willingness of artists to present their work in their place of both living and work, 'consumption of art in the artists' studio developed into a consumption of the studio too.'[34] This appropriation of the studio by the middle class meant that an inadvertent outcome of performance and conceptual art was 'the success of the studio' itself. The studio became a coveted model for metropolitan living. After artists had appropriated the spaces freed up by displaced industrial work, a new middle class in turn struggled for the same spaces, turning the ex-industrial into a booming property category. The studio loft has become a key urban trope, reappearing in modulated form in all subsequent cycles of post-industrial real estate boom and bust (1980s, 1990s and 2000s). On the other hand the gentrification process that ensued produced shared conditions between artists and low income residents in the fallout zones of capital's reoccupation of the inner city and this was generative of new struggles and solidarities around housing and space, notably in NYC's Lower East Side.[35]

---

32   Revisiting Macunias's projects in the long recession of the 2010s, Florian Cramer and Renee Ridgeway
      discuss how 'Fluxus was just as much an economic as it was an artistic project' and in Macunias's
      ambitious projects 'its performativity and processuality were not merely aesthetic but also biopolitical
      and bioeconomic choices that resulted in manic-depressive business cycles.'
      Florian Cramer, 'Depression: Post-Melancholia, Post-Fluxus, Post-Communist, Post-Capitalist, Post-
      Digital, Post-Prozac', in Maya Tounta ed., A Solid Injury to the Knees, Vilnius: Rupert, 2016, pp.60-107,
      p.87. Macunias may be known as the 'father of SoHo' i.e. gentrification, or one of a handful of artists
      who halted Robert Moses's expressway from destroying swathes of Lower Manhattan, or the progenitor
      of small but beautiful artists' cooperative housing projects thriving in post-crisis USA presently; what is
      clear is that his successes were not the ones he had intended:
      'Maciunas may be seen as an artist whose primary works were economic experiments, the lifelong
      endeavor of translating a communist concept of political (= macro) economy into viable micro-
      economies. Projects that boomed and busted, running in perpetual bipolar cycles of euphoria and
      depression. In this sense, Maciunas did not only pioneer gentrification but he also preempted the
      creative dotcom economy with its manic-depressive model of incubators and startups.' Cramer,
      'Depression: Post-Melancholia, Post-Fluxus, Post-Communist, Post-Capitalist, Post-Digital, Post-Prozac',
      p.92.
33   Zukin, Loft Living.
34   Zukin, Loft Living, p. 80.
35   Gregory Sholette et al (eds), Upfront: A Publication of Political Art Documentation/Distribution, 1983

Open or Community Studio (1966-Now)

Open or community studios took hold where the concentration of marginalized (classed and raced) people met with conditions of economic decline and urban dilapidation. A phenomenon known as spatial concentration whose crisis point was reached in the 1950s in western cities and which would quickly reverse into spatial deconcentration;[36] a term which we use here to describe complex migratory processes such as white middle class suburbanization in the US and, in the UK, the planned working class displacement from major cities to new towns.[37] The open studio registers and responds to the assault on and abandonment of urban communities left in the wake of these processes. In the US the Black Arts Movement built on the radical political premise that the 'ghetto itself is the gallery'[38] by establishing theatres, concert halls, rehearsal spaces, exhibition spaces, art and music studios in largely black and poor areas of the inner city.[39] The venues generated served as the platform for a rapidly developing ethos of community arts, characterized by the attempt to deflect specialist audiences in favor of direct and immediate community provision. In the UK such spaces and the community arts they supported 'dated back to the 1960s and were associated with alternative bookshops, theatre groups and the so-called Arts Laboratories that had succeeded in attracting new, younger audiences'.[40] But after an Arts Council report (1974) the logic of community arts was formalised and increasingly attracted fine artists and professional protagonists together with an expanding list of media artists.[41] These arts 'spaces' drew upon infrastructures and funding streams which were the legacy of the post-World War II welfare state (e.g. public health and community education), they were often open air, flexible, temporary or mobile, and more informal than the traditional studio or gallery.

Most community artists had a base or 'resource centre' for their operations (sometimes a mobile one) and employed a variety of facilities, media and techniques – dance, drama and writing classes, festivals, inflatables, murals, performance, photography, printing presses, play structures, video – which they used to foster public participation and to teach skills.[42]

---

and Neil Smith, *The New Urban Frontier: Gentrification and the Revanchist City*, London & New York: Routledge, 1996.

36   Yolanda Ward, 'Spatial Deconcentration in Washington D.C.' in Midnight Notes, *Space Notes – Midnight Notes*, No.4, 1981.

37   Peter Mandler, 'New Towns for Old', in B. Conekin, F. Mort C. and Waters, C. (eds), *Moments of Modernity: Reconstructing Britain, 1945-1964*, London; New York: Rivers Oram Press, 1999, pp. 208–27.

38   Emory Douglas quoted in Steven W. Thrasher, '"The ghetto is the gallery": black power and the artists who captured the soul of the struggle', 2017. [online] Available at: https://www.theguardian.com/artanddesign/2017/jul/09/ghetto-gallery-black-power-soul-of-a-nation-lorraine-ogrady-melvin-edwards-william-t-williams [Accessed 20 September, 2018].

39   Notably these were each focused on large experimental music ensembles formed by black artists. Two recent studies: George E. Lewis, *A Power Stronger Than Itself: The AACM and American Experimental Music*, Chicago: University of Chicago Press, 2008 and Benjamin Looker, *Point from Which Creation Begins: The Black Artists' Group of St. Louis*, St. Louis, MO: Missouri Historical Society Press, 2004, emphasise such community spaces in the development of radical black art.

40   John A. Walker, *Left Shift: Radical Art in 1970s Britain*, London: Tauris, 2002, p.130.

41   Walker, *Left Shift: Radical Art in 1970s Britain*.

42   Walker, *Left Shift: Radical Art in 1970s Britain*, p. 131.

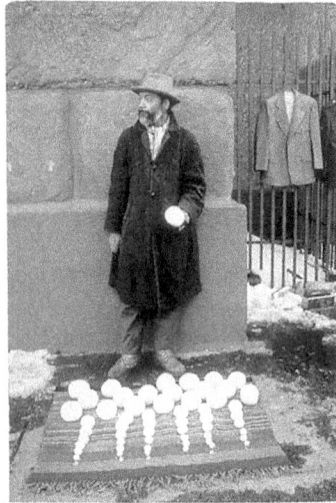

Fig.3: David Hammons, Bliz-aard Ball Sale, Cooper Square, New York, 1983. Photo by Dawoud Bey.[43]

In the UK, as the attempt to take art to the people faltered along with post-war optimism, the modus operandum of community arts increasingly fused with engaged and post-conceptual art taking it into the classroom (as in Central St. Martin's 'A' Class), the hospital (Loraine Leeson & Peter Dunn) as well as the factory and prison (Artist Placement Group).[44] Later, during the 1980s, a period of austerity, cutbacks to welfare provision and state retrenchment, the same inner-city areas where community arts had become integrated into local state service provision and funding became sites of spatial deconcentration, i.e. the breaking up of the ghetto and gentrification of working class areas.[45]

In Africa, group practices such as Laboratoire AGIT Art in Senegal and Huit Facettes established community studio projects in both urban and rural situations as responses to the harsher social climate and degenerating prospects for critical art created by Structural Adjustment Programmes.[46] The community or open studio tended to embed itself within sites of social reproduction, but as that is disaggregated, then to respond to and encompass a situation of actual or perceived spatial dispersion. Whilst poor or poorly maintained housing would be a central theme throughout this moment, the exhibition, display, performance or presentation of art tends towards the street or local public realm. In the West, as working class people began to be displaced from areas where community resources had helped develop

43   *David Hammons - Blizaard Ball Sale* (1983) by Cea. is licensed with CC BY 2.0. To view a copy of this license, visit https://creativecommons.org/licenses/by/2.0/

44   See Walker, *Left Shift: Radical Art in 1970s Britain.* and Marina Vishmidt, 'Creation Myth' [online] Available at: http://www.metamute.org/editorial/articles/creation-myth [Accessed 20 September, 2018].

45   Yolanda Ward. 'Spatial Deconcentration in Washington D.C.' in Midnight Notes, *Space Notes – Midnight Notes*, No.4, 1981.

46   Okwui Enwezor, 'The Production of Social Space as Artwork, Protocols of Community in the Work of Le Groupe Amos and Huit Facettes', in Gregory Sholette and Blake Stimson (eds.), *Collectivism After Modernism*, Minneapolis: University of Minnesota Press, 2007, pp. 223–252.

their autonomy (buildings, squats, social centers), community arts took to the street, often framing the struggles for self-determination and the bathos of economic survival on the breadline as resistant creativity.

The community studio's spirit of improvisation and spatial repurposing was gradually incorporated into the governmental push for a non-stop programme of visitor-friendly arts festivals whose presence marks a wider geopolitical competition over place making, cultural tourism and inward investment. For those communities who initiated these participatory practices, the managerial turn in cultural commissioning replicates a more widespread shift in the nature of power which claims to speak in the name of the voiceless, vapidly invoking communities while divesting them as a source of meaning making from their own representations.[47] The authentic community studio lives and dies with the waning of universal provision and in inverse relation to the rise of built space as high rent-yielding private property. With declining social tenure and public ownership tracking an exponential rise of community arts, the community studio has all but disappeared, retreating to bedroom production and the corporate ghettos of social media.

Fig. 4: Makrolab mkII, Rottnest Island /Wadjemup, Western Australia, 2000, Part of the Home exhibition, Art Gallery of Western Australia and Perth International Festival. Photo by Marko Peljhan.

The Networked Studio (1989–present)

The administrative and informational aesthetics pioneered by 1960s factory-based conceptual artists and inspired by cybernetics and communication technologies, would reemerge in the ubiquitous office metaphor that was incorporated into the computer desktop and popularized

---

47   Walker, 2002, op. cit. and Claire Bishop, *Artificial Hells: Participatory Art and the Politics of Spectatorship*, London: Verso. 2012.

by the Apple Macintosh from 1984.[48] Turing's universal machine was to produce a new spatial topology in which art can pass as – and be mistaken for – generic labor within the cybernetic society. To the casual observer, computer use by artists is largely indistinguishable from that of a service worker, journalist, architect or engineer. The conceptual techniques of the 1960s were eventually reassembled and compacted in the 1990s network studio and the figure of the immaterial laborer.[49] The isolated studio does not necessarily move position but becomes increasingly networked by communications technologies, services, changing terms of building insurance, health & safety specifications, and the further development of financial claims which feed the growth of tertiary industries and bureaucracies. The artist in the networked studio responds to the limits breached by their precursors who collapsed the divisions active in wider society and, in the first wave, overlapped significantly with the DIY community spirit of the open studio artist. The network studio camouflages itself in a number of guises be that the office, the artist's loft, the hacklab, social center, cybercafé or simply the portable laptop. We can therefore position the network studio as having an uncertain, deceptive and possibly anxious relationship to built space. This can be connected to a double movement, identifiable within its brief history, of artists occupying spaces that advertise their networked infrastructures and those that dissemble them within the shell of the old city.

The networked studio in its most extreme form is the mobile home that the infonaut carries with them through their connection to the network. The early network artist affirmed spatial precarity as digital nomadism like it was a blessing. This 1990s affirmation of the network as a replacement of fixed spatial relations tended towards the production of singular, albeit hybrid spaces (eg. London's Backspace, Ljubljana's Ljudmila, New York's Rhizome, and Zagreb's MaMa), in which artists built assemblages of hard- and software to explore and magnify the dawning universal space/time of the internet. The network studio provides a place to tune into and amplify an increasingly chaotic out there. The most iconic example of this was perhaps the Makrolab (1994-2007), devised by the Slovenian artist Marko Peljhan, who had grown up in non-aligned Yugoslavia listening to CB radio and transposed the experience of remote and isolated listening to the noise of the information age.[50] The Makrolab was a mobile communications monitoring hub that looked like a space-pod and moved between ever more isolated places gathering the public and monitoring the often secret or invisible telecommunications exchanges of incipient global capitalism.

---

48   Benjamin Buchloh, 'Conceptual Art 1962-1969: From the Aesthetic of Administration to the Critique of Institution', *October*, No.55, 1990, pp. 105-143.

49   Maurizio Lazarrato, 'Immaterial Labour', in Paolo Virno and Michael Hardt (eds), *Radical Thought in Italy: a Potential Politics*. Minneapolis, Minn; London: University of Minnesota Press (Theory Out of Bounds), 1996.

50   'One of the current projects I am doing for Documenta is called Makrolab. It is a project that will research isolation strategies: how to isolate oneself from society to reflect and see this society better. It is an opposite of the usual going into the society and trying to change or make things. My thesis is that a small amount of people in an isolated and insulated environment with completely open possibilities of communication and monitoring of social events, but physically isolated, can provide a much faster, further and more efficient "call" (?) for social evolution. It is my thesis, not just an idea, and I am going to prove it.' Peljhan quoted in Josephine Bosma, 'Interview with Marko Peljhan', 2011, available: http://www.josephinebosma.com/web/node/61 [Accessed, 11th September 2018]

This mobility, however, also presages the impending spatial precarity of the millennium in which to survive is to keep moving on, to 'replace [...] the inevitability of being uprooted with the strategy of pitching and breaking camp', as the urban stickering campaign of artist-collective Inventory had it (Inventory, 1999–2002). It is striking that the isolation Makrolab achieved was underwritten by the same forces of globalization that its periscopes peered at; not only satellites, cables, deep dishes, antennas and data flows but, indirectly at least, the transnational investment strategies of George Soros and his Open Society Institute.[51] Thus, the networked utopia of the artist-infonaut dovetailed beautifully with the migratory practices that would be the prerequisite of globalization's 'race to the bottom', and in this sense the mobile isolation of the studio can be seen as an analogue to the precarious and alienated status of the worker within the new world order, the so-called digital nomad.[52]

The appearance of the digital avant-garde also coincides with the advent of the last great real estate boom, (in the UK and US), of which we are yet to see the end. In this moment artists, often collectively, obtained workspace in the inner city if only for a brief while. These digital artisans were often artists who moonlighted as web designers and could be found producing artistic and commercial work on the same machine at the same desk. To the office workers next door, their activity was indistinguishable from white-collar work. While at first the network studio had projected itself as an anomaly amidst the commercial zones of the city, this possibility was rapidly eradicated by rising rents based upon projections of burgeoning commercial demand. The network studio was soon to give way to businesses that understood themselves as creative in themselves. Characteristic of the new format of businesses moving into fill these spaces were built-in signifiers of leisure (ping pong tables, bean bags, office pets, drinks trolleys, beer taps, house plants) incorporated into the new corporate aesthetic. The technology which denizens of the networked studio had experimented with swiftly began to reconfigure the space between previously separated spheres, not least labor and creativity. The playful 'hacking' of the social outcomes of computer technologies, now incorporated into a host of new business models, gave way to a calculative logic bearing down on what had previously remained uncalculated, profoundly disrupting the social through understanding it as a field available for continuous commercial speculation. Now profits could be sought through the elimination of space/time between those previously discreet social forms, objects, sites and resources within a 'becoming topological of culture' in which any data point can be connected to any other.[53] The integration of cybernetics with a competitive and reifying economic logic (sorting, ranking, data-banking, locating, connecting) unleashes a ubiquitous entrepreneurialism based on processes of disintermediation. Through disintermediation, points in space, along with buyers and sellers, can be connected in new ways that undermine older spatio-temporal exclusivities. This opening of the social field to disruptive reconfigurations fed into the emergence, in the late 1990s, of an entrepreneurial subject constantly on the look-out for commercial opportunities (and venture capital) within

51   The Open Foundation attempted the cultural integration of eastern with western Europe invaluable to the creation of the *smooth space* required by the transnational movement of capital (Republic of Slovenia, Ministry of Culture, nd.).
52   Tsugio Makimoto & David Manners, *Digital Nomad*, New York: Wiley, 1997.
53   C. Lury, Luciana Parisi, Tiziana Terranova, 'Introduction: The Becoming Topological of Culture', *Theory, Culture & Society*, Vol. 29, No 4-5, 2012.

new and unstable proximities, be that physical, resource-based or both.[54] Within this new paradigm, the operator of the network studio, who had up to this point sought to dissolve their distinction in the hybrid commercial-cultural spaces of the city and the internet,[55] made a counter-movement, seeking instead, against the creeping creative branding of the dot com economy to claim the distinction of art and the activity of the artist. The networked studio began to disappear as all studios became digitally networked-by-default. The chaos that deregulated transnational capitalism unleashed in the form of financialization and labour force restructuring would be converted, at an urban scale, into the blow-down of social housing estates and the vertical eruption of steel and glass office blocks in which there is no longer hope of stable and affordable studio space. The networked studio can be seen as the early warning signal for the studio's permanent ungrounding.

Fig. 5 The Artworks Pop Up container park, Elephant & Castle, privately run but heavily sponsored by developer Lendlease, photo from September 2014.

The Pop-up Studio (2000–2020)

The post-conceptual artist doesn't need a studio just as much as they can no longer have a studio. Nonetheless the studio picks itself up and walks on its own two feet, apparently popping-up in new and pre-fabricated pseudo-public spaces. It is a creature of municipally-led regeneration schemes that imitates and fuses with a prior moment of artist's self-organized exploitation of tenancy breaks in commercial properties, such as Tracy Emin and Sarah Lucas'

---

54   AirBnB has come to epitomise such a topologisation of culture and the networking of social and physical relationships which intensifies a reifying logic throughout the social field.

55   Net artists Heath Bunting and Rachel Baker, for instance, liked to work in Easyeverything cafés among casual surfers: 'I like it there', Bunting has remarked, 'as most people are checking hotmail while I am doing programming – it's a good disguise.' Josephine Berry, The Thematics of Site-Specific Art on the Net, PhD Thesis, University of Manchester, [online] www.metamute.org/sites/www.metamute.org/files/thesis_final_0.doc 2001, 81)

The Shop on Brick Lane (1993). By this point, the isolated studio is not just surrounded by the networks of urban financialization, but directly integrated into their centers of development.

The pop-up studio is now a developer-led must-have for any land subject to speculative investment or seeking it. This speculative function of the studio is symptomized in the highly visible nature of these spaces of exception in which, regardless of actual requirements, glazed shop-fronts are installed, flashy signage and bright colors deployed, and artists expected to charismatically perform like artists no matter how dryly conceptual, computer based, post-studio or socially engaged their practice might be. In actuality, the pop-up studio is so integrated into the endo-colonial process of urban transformation that it is more isolated from both the artist's own agency and the social than ever. It comprises a significant push for art's role in incubating obsolescent spaces driven by both small arts organizations (or the microentrepreneurs steering them) and government policy. Pop-ups, meanwhile spaces, space ache, pop-up retail, interim spaces, slack spaces emerge as a new language directly after the 2007/8 financial crisis. The group Meanwhile Space pioneered the mediation between empty spaces, money-saving councils and investors on the prowl.[56] 'Empty spaces are a blight to communities, a financial drain to owners and stimulate wider civic problems. To us they are an opportunity'.[57] The pop-up has a specific temporality, its presuppositions are set by the speculative proposition that empty space will become realizable value and in turn it assists this prediction into becoming reality. Realizing the age-old logic of crisis as an opportunity, the pop-up also operates through art-without-qualities to arrive at the consensus that anything can pop-up but, as the rents rise, anything increasingly tends to look like some form of commerce. This may be the first time that the studio as a form fully precedes, presupposes and overcodes the activities of artists.

The pop-up is shot through with contingencies, requirements and conditions: a fixed creativity ratio prescribed in the developer's plan with the space allocated through competition. Its exceptionality provides the necessary association with excellence its commercial sponsors require.[58] Practice in the pop-up studio is transparent like a fishbowl, intended to be viewed, visited and displayed. A space in which artistic labor is as performative as it is absorptive. Whether it's filled with cupcakes or high-end abstraction this will be a conspicuously ethical performance, in which the performer fails to notice the conditionality their striving imposes on everyone else. What masquerades as inclusivity is in fact the conduit to exclusivity. But, given there is no specifically identifiable characteristic of pop-up art, who is the pop-up practitioner?[59]

56   Mara Ferreri, *Occupying Vacant Spaces: Precarious Politics of Temporary Urban Reuse*, PhD Thesis, Queen Mary University of London, 2013, [online] Available at: <http://qmro.qmul.ac.uk/jspui/handle/123456789/8460> [Accessed, 28/09/2018]i, 2013).
57   Meanwhile Space quoted in Ferreri, *Occupying Vacant Spaces*, p.129.
58   An example of this is the Alumno / SPACE Studios / Goldsmiths University of London bursary in which 'one graduate receives one years [sic] rent free studio space in the artists [sic] studios Alumno developed and which are operated by SPACE in the former Southwark Town Hall Building which also houses Goldsmiths students.' http://alumnogroup.com/alumno-space-studios-goldsmiths-bursary-winner-2019/.
59   The Pop Up People report published in February 2012 by the Empty Shop Network, for example,

The pop-up studio is therefore: 'a very vexatious thing full of metaphorical subtlety and theological perversity'.[60] Behind a smokescreen of community it puts a young, precarious, flexible demographic to work as quisling managers, 'trusted middle persons'.[61] Their agency from the point of view of local government is that they stage a public which isn't the public local government is usually bound to provide services to. This trusted middle mediates interests which are closer to those of developers and international investors than those of artists or the community in whose name the spaces are contrived.[62] The pop-up people are then the front of house and community liaison for a back room that is selling off the entire house.

Now that pop-up retail is the norm, it is easy to forget the form's origins as a local authority strategy of allocating properties managed by them to temporary housing or community use. Throughout the 1970s short life housing had been both a means of alleviating housing need, and meeting demands for space for a wide range of emergent community needs in the face of local authorities' squandering of empty and decaying public property.[63] The pop-up has represented some sort of survival of the studio situation for artists and therefore also the short-life strategy, (even for those who end up refusing the instrumental roles imposed upon them in situ).[64] Through the intensification of property guardianship schemes after the 2008 property market crash many artists obtained live/work studios in former local authority housing. It is in just such a situation that photographer, Rab Harling, property guardian for Bow Arts Trust, turned his lens on the estate and his activist proclivities towards the peculiar public-private arrangements which had led to the eviction of both tenants and leaseholders previously living on the estate. At the Balfron Tower, artists and residents worked together to expose and block two particularly offensive public art works synonymous with the excesses of public art in regeneration.[65] Public art commissioning agency, Artangel, who site new major

offers a depiction of the personal and professional characteristics required to become a pop-up shop practitioner, and begins with the line "Pop Up People: are truly entrepreneurial, even if their project is more about community than commerce".' Thompson, quoted in Ferreri, op. cit., 2013, p.133.

60   Karl Marx, *Capital: A Critique of Political Economy Vol.*I, Trans. Ben Fowkes and David Fernbach. London: Penguin, 1990.

61   Ferreri, *Occupying Vacant Spaces*, p.133.

62   In examining failures in democratic representation, it is also relevant to look at the professional background of elected members themselves. According to Paul Watt and Minton, in 2013 just under 20% of Southwark's 63 councillors worked as lobbyists, while the former leader of the council, Jeremy Fraser, went on to found lobbying firm Four Communications, where he was joined by former councillor and cabinet member for regeneration Steve Lancashire. Equally concerning is the 'revolving door' between council employees and elected members and developers. For example, Tom Branton was Southwark Council's project manager for the Elephant & Castle Regeneration Project until 2011, when he left the Council to work for Lend Lease, the lead developer on the project' See: Paul Watt and Anna Minton, 'London's Housing Crisis and its Activisms', *City*, 2016, available at: https://www.tandfonline.com/doi/abs/10.1080/13604813.2016.1151707 [Accessed: 15 July 2019].

63   Anna Bowman, *Interim Spaces: Reshaping London – The Role of Short Life 1970-2000.',* 2010, available at: https://research-information.bristol.ac.uk/files/34489446/289648.pdf [Accessed, 10 March 2019], pp. 99-100.

64   Richard Whitby, 'Angels, The Phoenix, Bats, Battery Hens and Vultures – The Bow Arts Trust Live/Work Scheme', self-published 2011, https://www.academia.edu/2560146/Angels_The_Phoenix_Bats_Battery_Hens_and_Vultures_The_Bow_Arts_Trust_Live_Work_Scheme [Accessed: 15 July 2019].

65   Caroline Christie, 'Hey Creatives, Stop Fetishising Estates', *Vice*, 14 July, 2014, available at: https://www.vice.com/en_uk/article/jm953d/balfron-tower-art-fetishising-estates-157 [Accessed: 15th July 2019]

commissions in edgy locations, may be considered pioneers of pop-up art which parasites off and re-encodes blighted spaces. The resulting temporary artworks often function to rebrand an area or iconic building by socializing the site differently, drawing the attention of those deemed most likely to invest in its future. With artists often working in situ, such commissioning approaches have arguably prepared the ground for the pop-up studio sited temporarily in the midst of development zones, whose users are encouraged to make work as visibly as possible and regularly offered ad-hoc exhibition opportunities until the available space dries up.

Viewed retrospectively, after the cycle beginning with the financial crisis of 2007-2008, the pop-up was a moment in these former places' violent transport onto the open market. Striking here is how it is now the council flat or residential estate that has become newly designated as a (temporary) site of studio production or artistic intervention. In the same way that industrial spaces were planned out of the city in the 1960s, curtailing their actual utility, financialized urban development's need for new sites of value extraction chaotically inflicts long-term obsolescence on entire communities and their homes producing only short-lived opportunities for artists. This constitutes a total inversion of avant-garde modernism's dreams of masterplanning the city, with artists no longer employed as visionaries but rather encouraged to pick over the bones of social housing provision.

The house, whether in the form of the evacuated single terraced British House (Rachel Whiteread, 1993) or the housing estate maps refined through painterly abstraction, as Estate Maps (Keith Coventry, 1991-1995), became a central motif of art in the UK at a moment when the model of habitation seemed at once connected to geopolitical change and reflective of the waning of artistic modernism and a microcosm of wider social antagonisms. If 1990s houses in art were ciphers for a mourning of the past or for socialism, the late-2000s placement of artists in former local authority housing would produce critical meditations on the image of the house and the social question of housing (Jessie Brennan, 2015-ongoing and Laura Oldfield Ford, 2008-ongoing) as well as some very public moments of hubris as artistic ambition met with raw discontent over housing poverty and the mistreatment of tenants facing estate regeneration (Mike Nelson, 2013). The small number of artists who revolted against these grotesque conditions from inside them to expose exactly how they were not inevitable or natural, were making and inventing politics where they had been effaced as much as a critical art where a feebler sibling was intended to emerge.[66] Looking back from 2009 to the present, this is one of the few perspectives from which housing privatization, art, space and property enter into a fully antagonistic and dynamic entanglement by which art's critical relationship to society is exercised and developed.

and Hannah Ellis-Petersen, 'Decaying East London Tower Block to House 12-hour Macbeth Production', *The Guardian*, 19 June, 2014, available at: https://www.theguardian.com/stage/2014/jun/19/east-london-balfron-tower-macbeth-production [Accessed: 15 July 2019].
66   Christy Romer, 'Artist squares up to Regulator over "manifestly unreasonable" fundraising investigation', *Arts Professional*, 2017, available at: https://www.artsprofessional.co.uk/news/exclusive-artist-squares-regulator-over-manifestly-unreasonable-fundraising-investigation [Accessed: 20 September 2018].

Conclusions

If, throughout modernity, succeeding generations of artists dreamed of accelerating both thought and creation to the tempo of technological production they could not but fail. The paintbrush's pursuit of fugitivity moved at a slower pace than the steam age, while the Futurists' ecstatic white-heat left the terrestrial assembly line in the dust. Temporal non-coincidence with prosaic production has been inherent to art, whether intentionally or not, for centuries. Yet the artist's requirement to pop-up, fill-in, engage and performatively be there, evidences an inescapable synchrony with the accelerated and aimless cycles of creation and destruction, bubble and burst, borne of financialization.

That Mayor Sadiq Khan's recent London Plan calls for the provision of permanent spaces for artists is more an indication of the problem than a sign of its solution.[67] The pecuniary provision of such spaces echoes the exceptionality that art still monopolizes while in practice poisoning it at the roots, since it entails the creation of special protections for existing studio spaces and thereby imposes requirements on art to do something in return for its permission to exist. The studio as the space of a practice distanced from the rest of life seems to be facing extinction, either because it's unaffordable, it now lies too far from home, or its occasional possibility of fulfilling the terms of its conditions overwhelms practice altogether. But if the gulf between an earlier womb-like isolation and today's developer-fantasised spectacular creative performances seems too great to retain art as their common denominator or outcome, there is inevitably a secession from both these models that is where whatever might actually feel like art is taking place. This truth is as likely to entail the total obsolescence of the, in historical terms, relatively short-lived model of autonomous art as to relocate it safely elsewhere. Capitalism's unending production and destruction of space has mutated the physical auspices of art's production, and those spatial evolutions have likewise driven a transformation of artistic practices and vice versa. The inhabitation of available spaces or conduits of production can be said to condition the dialectic of art's relationship to the social world as well as artists' opportunistic infestation of niches of survival. That this process has itself been converted into the paradigm and lubricant of spatial prospecting may finally mean that, as with the burning of the earth's carbon resources, like life new art is ever more unsustainable. Yet as we are starting to see, the cannibalization of art by capitalism engenders encounters between vagrant aesthetics and planetary dispossession that cannot but unite against the very forces that condemn them to a plodding nomadism or show pony servitude.

References

Adorno, Theodor W. *Aesthetic Theory*, trans. Robert Hullot-Kentor, London; New York: Continuum Press, 1997.

Arts Council England, Art in Empty spaces: Turning Empty Spaces into Creative Spaces Grant, 2009, available at <http://www.artscouncil.org.uk/about- 324us/investment-in-arts/action-recession/art-empty-spaces/> [Accessed 2 March February 2011].

Balzac, Honoré De. 'The Unknown Masterpiece', 1845, available at: https://www.gutenberg.org/files/23060/23060-h/23060-h.htm#link2H_4_0002.

---

67   https://www.london.gov.uk/sites/default/files/draft_london_plan_chapter_7.pdf.

Baumgärtel, Tilman. 'I Don't Believe in Self-Expression: An Interview with Alexei Shulgin', 1997, available at: <http://www.intelligentagent.com/archive/fall_shulgin.html> [Accessed, 5 September 2018].

Beech, Dave. *Art and Value: Art's Economic Exceptionalism in Classical, Neoclassical and Marxist Economics,* Leiden: Brill, 2015.

Berry, Josephine. *The Thematics of Site-Specific Art on the Net*, PhD Thesis, University of Manchester, 2001, available at: www.metamute.org/sites/www.metamute.org/files/thesis_final_0.doc [Accessed, 3rd June 2021].

Berry Slater, Josephine and Iles, Anthony. *No Room to Move: Radical Art and the Regenerate City*, London: Mute Books, 2010.

Bishop, Claire. *Artificial Hells: Participatory Art and the Politics of Spectatorship*, London: Verso, 2012.

Bosma, Josephine. 'Interview with Marko Peljhan', 2011, available: http://www.josephinebosma.com/web/node/61 [Accessed, 11th September 2018]

Bowman, Anna. *Interim Spaces: Reshaping London – The Role of Short Life 1970-2000.',* 2010, Available at: <https://research-information.bristol.ac.uk/files/34489446/289648.pdf [Accessed, 10th March 2019].

Buchloh, Benjamin H. 'Conceptual Art 1962-1969: From the Aesthetic of Administration to the Critique of Institution', *October*, 55, 1990, pp. 105–143.

Buren, Daniel. 'The Function of the Studio', *October*, Vol. 10, 1979, pp. 51–58.

Bryan-Wilson, Julia. *Art Workers: Radical Practice in the Vietnam War Era*. Berkeley: University of California Press, 2009.

Christie, Caroline. 'Hey Creatives, Stop Fetishising Estates', *Vice*, 14 July, 2014, available at: *https://www.vice.com/en_uk/article/jm953d/balfron-tower-art-fetishising-estates-157*

Conlin, Peter. *Slack, Taut and Snap: A Report on the Radical Incursions Symposium | Mute*, 2009, available at: http://www.metamute.org/editorial/articles/slack-taut-and-snap-report-radical-incursions-symposium [Accessed: 15 July 2019].

Clark, T.J. *The Painting of Modern Life: Paris in the Art of Manet and his Followers*, Princeton University Press, 1999.

Cramer, Florian. 'Depression: Post-Melancholia, Post-Fluxus, Post-Communist, Post-Capitalist, Post-Digital, Post-Prozac', in Maya Tounta (ed) *A Solid Injury to the Knees*, Vilnius: Rupert, 2016, pp. 60–107.

Ellis-Petersen, Hannah. 'Decaying East London Tower Block to House 12-hour Macbeth Production', *The Guardian*, 19 June, 2014, available at: https://www.theguardian.com/stage/2014/jun/19/east-london-balfron-tower-macbeth-production [Accessed: 15 July 2019].

Enwezor, Okwui. 'The Production of Social Space as Artwork, Protocols of Community in the Work of Le Groupe Amos and Huit Facettes', in Gregory Sholette and Blake Stimson (eds) *Collectivism After Modernism*, Minneapolis: University of Minnesota Press, 2007, pp. 223–252.

Foucault, Michel. 'Nietzsche, Genealogy, History', in Paul Rabinow (ed), *The Foucault Reader*, London: Penguin Books, 1984.

Ferreri, Mara. *Occupying Vacant Spaces: Precarious Politics of Temporary Urban Reuse*, PhD Thesis, Queen Mary University of London, 2013, available at: https://core.ac.uk/download/pdf/77038644.pdf [Accessed June 3 2021].

Goldner, Lauren. 'Fictitious Capital for Beginners', *Mute,* Vol.2 No.6, 2007.

Gould, Charolotte. *Artangel and Financing British Art: Adapting to Social and Economic Change*, London & New York: Routledge, 2018.

Harding, Anna (ed), *Artists in the City: SPACE in '68 and Beyond*, London: Space Studios, 2018.

Harris, Andrew. 'Livingstone versus Serota: The High-Rise Battle of Bankside'. *The London Journal*. Vol. 33, No. 3, 2008, pp. 289–99.

Harvey, David. 'The Spatial Fix: Hegel, von Thunen and Marx'. *Antipode* Vol.13, no. 3, 1981, pp.1–12.

Harvey, David. 'Globalization and the "Spatial Fix"', *Geographische Revue*, No.2, 2001, pp. 23–30

Harvey, David. 'The Art of Rent', *Socialist Register*, 2002, pp. 93–102.

Hayward, Danny. 'Fire in a Bubble', *Mute*. 15th September 2017, available at: http://www.metamute.org/editorial/articles/fire-bubble [Accessed 20 August, 2018].

Inventory, *Inventory Sticker Project*, c.1999-2002. A complete collection can be consulted in May Day Rooms Archive, London, <https://maydayrooms.org/>

Iles, Anthony & Vishmidt, Marina. 'Make Whichever You Find Work'. *Variant*. No. 41, 2011.

Jacobs, Jane. *The Death and Life of Great American Cities*, New York, Random House, (1961) 1993.

Kant, Immanuel. *Critique of Judgment*, trans. Werner S. Pluhar, Indianapolis: Hackett, 1987.

Lapavitsas, Costas. 'Financialisation, or the Search for Profits in the Sphere of Circulation', London: SOAS, 2009, available at: http://www.soas.ac.uk/rmf/papers/file51263.pdf, [Accessed 20th August, 2018].

Lazarrato, Maurizio. 'Immaterial Labour', in Paolo Virno and Michael Hardt, (eds), *Radical Thought in Italy: a Potential Politics*. Minneapolis, Minn; London: University of Minnesota Press (Theory Out of Bounds), 1996.

Lefebvre, Henri. *The Survival of Capitalism*, trans. F. Bryant, London: Allison & Busby; St Martin's Press, 1976.

Lefebvre, Henri. *The Production of Space*, trans. D. Nicholson-Smith, Oxford; Cambridge, Mass.: Blackwell, 1991.

Lewis, George E. *A Power Stronger Than Itself: The AACM and American Experimental Music*, Chicago: University of Chicago Press, 2008.

Looker, Benjamin. *Point from Which Creation Begins: The Black Artists' Group of St. Louis*, St. Louis, MO: Missouri Historical Society Press: Distributed by University of Missouri Press, 2004.

Lury, Celia, Parisi, Luciana & Terranova, Tiziana. 'Introduction: The Becoming Topological of Culture', *Theory, Culture & Society*, 29 (2012): 4–5.

Makimoto, Tsugio & Manners, David. *Digital Nomad*, New York: Wiley, 1997.

Mandler, Peter. 'New Towns for Old', in B. Conekin, F. Mort C. and Waters, C. (eds.), *Moments of Modernity: Reconstructing Britain, 1945-1964*, London; New York: Rivers Oram Press, 1999, pp. 208–27.

Martin, Randy. 'Introduction', in *Wohnungsfrage*, Berlin: Haus Der Kulturen Der Welt. Exhibition 22nd October – 14th December 2015, (n.p.).

Marx, Karl. *Capital: A Critique of Political Economy Vol. I*, trans. Ben Fowkes and David Fernbach, London: Penguin, 1990.

Medina, Cuauhtémoc. 'The "Kulturbolschewiken" II: Fluxus, Khrushchev, and the "Concretist Society"', *RES: Anthropology and Aesthetics*, No. 49/50 (Spring – Autumn, 2006), pp. 31–243.

Midnight Notes, 'Postscript: Space, and Race Space', *Midnight Notes*, No.4, 1981, pp. 32-36.

Miller, Toby. & Shin Joung Yeo, 'Artists in Tech Cities', London: Space Studios, 2017, available at: <https://repository.lboro.ac.uk/articles/report/Artists_in_tech_cities/9466355>, [Accessed, 3 June 2020].

Moreno, Louis.'The Urban Process Under Financialised Capitalism', *City*, 2014, 18:3, pp. 244-268.

Moorhead, Joanna. 'Artists are coming to a high street near you', 23 April 2009, available at: https://www.theguardian.com/artanddesign/2009/apr/23/artists-take-over-empty-shops [Accessed, 20 September 2018].

Romer, Christy. 'Artist squares up to Regulator over "manifestly unreasonable" fundraising investigation', *Arts Professional*, 2017, available at: https://www.artsprofessional.co.uk/news/exclusive-artist-squares-regulator-over-manifestly-unreasonable-fundraising-investigation [Accessed, 20 September 2018].

Republic of Slovenia, Ministry of Culture. New Media Art Timeline. [online] Available at: https://www.culture.si/en/New_media_art_timeline [Accessed 13 February 2010].

Saler, Michael T. *The Avant-Garde in Interwar England: Medieval Modernism and the London Underground*. New York: Oxford University Press, 2001.

Sholette, Gregory. et al (eds) *Upfront: A Publication of Political Art Documentation/Distribution*, 1983.

Smith, Neil. *The New Urban Frontier: Gentrification and the Revanchist City*, London & New York: Routledge, 1996.

Sparrow, Andrew. '100 quangos abolished in cost-cutting bonfire', 2012. [online] available, https://www.theguardian.com/politics/2012/aug/22/100-quangos-abolished-bonfire [Accessed, 28 September 2018].

Thompson, Dan. 'Pop Up People', 2012, available at: https://emptyshops.files.wordpress.com/2012/06/popuppeoplereport.pdf [Accessed, 28 September 2018].

Thrasher, Steven W. '"The ghetto is the gallery": Black power and the artists who captured the soul of the struggle', 2017. Available at: https://www.theguardian.com/artanddesign/2017/jul/09/ghetto-gallery-black-power-soul-of-a-nation-lorraine-ogrady-melvin-edwards-william-t-williams [Accessed 20 September, 2018].

Vishmidt, Marina. 'Creation Myth', 2010. Available at: http://www.metamute.org/editorial/articles/creation-myth [Accessed 20 September, 2018].

Walker, John A. . *Left Shift: Radical Art in 1970s Britain*, London: Tauris, 2002.
Ward, Yolanda. 'Spatial Deconcentration in Washington D.C.' in Midnight Notes, *Space Notes – Midnight Notes*, No.4, 1981.

Watt, Paul and Anna Minton, 'London's Housing Crisis and its Activisms', *City*, 2016. Available at: https://www.tandfonline.com/doi/abs/10.1080/13604813.2016.1151707 [Accessed: 15 July 2019].

Whitby, Richard. 'Angels, The Phoenix, Bats, Battery Hens and Vultures – The Bow Arts Trust Live/Work Scheme', self-published 2011, https://www.academia.edu/2560146/Angels_The_Phoenix_Bats_Battery_Hens_and_Vultures_The_Bow_Arts_Trust_Live_Work_Scheme [Accessed: 15 July 2019].

Zukin, Sharon. *Loft Living: Culture and Capital in Urban Change*, Baltimore: London: The Johns Hopkins University Press, 1982.

MOURNING MACHINE: OBITUARY FOR A VANISHED PLACE

SYLVI KRETZSCHMAR

Have a look around, all gone

All the people are gone

There used to be all these drop-dead shops and bars

Interviews with former occupants, tenants, and contractors of the so-called Esso-Houses, which were originally located on Hamburg's Reeperbahn in the St. Pauli neighborhood, were set to music by twelve women with megaphones. The buildings had been purchased by the group Bayerische Hausbau in 2009, with the intention of demolishing them. The Esso Houses have become a significant part of St. Pauli's life, with their inexpensive homes, stores, and famed nightclubs and pubs. The houses were named after a gas station that served as the district's village square and had a 24-hour convenience store. It was once known as the heart of St. Pauli. The Esso Houses were evacuated in 2013 and demolished in 2014 due to an alleged imminent danger of collapse. The evacuation of the structures, as well as the eviction of the neighborhood's older and more established residents, has come to symbolize the interdependence of property speculation and urban development policy. The Megaphone Choir took part in protests against the St. Pauli district's growing gentrification. It has contributed to the political discourse surrounding the Esso Houses over the years.[1]

Um, well...

Of course you cannot stop it

Of course you cannot stop it

Of course you cannot stop it

Of course you cannot stop it

Of course you cannot stop it

---

1 The Megaphone Choir was a part of two PhD programs (Assemblies and Participation and Performing Citizenship) that pioneered the *Hamburg School* of Participatory Art Based Research (PABR) techniques between 2012 and 2017. PABR differs from academic research. PABR 'explored forms and formats of research in between art, academia and society that were meant to include not only artists and researchers but also members of other communities, such as kids, neighbours, activists, experts, citizens and non-citizens'. Together with activists and artists I publicly invented and tested the Megaphone Choir as a multiply movable space occupying voice amplification system. See: https://pab-research.de/pabr/.

Of course you cannot stop it

Of course you cannot stop it

Of course you cannot stop it

Fig. 1: Sylvi Kretzschmar/Megaphone Choir: Performance 2013 Park Fiction Hamburg, St. Pauli, Photo by Rasande Tyskar.

Of course you cannot stop Gentrification

But you can eke out

Some small islands

And above all

You can get on their nerves

You can get on their nerves a lot

And that's good somehow

Not to leave it to them

Just like that

But to drive them

A little bit crazy

I'm in the mood

To drive them really crazy

Although actually I'm not that kind of person at all

The Megaphone Choir as an Amplifier[2]

Megaphones have a pistol grip. The speaker uses it to target people and spaces, to address the speech in aligned acoustics. You need to pull a trigger while you are talking in order to shoot an oration. The megaphone as an object might be emblematic of political resistance, protest and civil rights movements but at the same time it is an apparatus of command and instruction, an instrument or gadget made to give orders and to fill the air with sounds of warnings and directives. It is used by the military, the police, safeguarding services, fire departments, commanders, commandants, coaches and (cheer)-leaders. It's meant for crowd management, emergency management, used to instruct, to teach or to win over a crowd. In all these areas of reference, the applications and functions of the megaphone oscillate in sound as well as in the image of the megaphone when it amplifies political messages — for example, in the street during a political demonstration. Against this background, it may not be that obvious but the Megaphone Choir arose from the question: What could a political speech be like if it was emerging from the political process, instead of initiating it, directing or controlling it? How might a political speech, rather than planning, specifying, or commanding activist action, effectively reflect and amplify a collective of activist action?

At its core, the Megaphone Choir is about transporting the statements from one place/ moment/context to another. Everything we speak, call, whisper and sing is the verbatim reproduction of interview answers. A choral repetition of the statements, based on the audio interviews, is designed and rehearsed. Like every choral speech, this calls for an intensification of the musical qualities of language. The individual speech melodies and speech rhythms of the interviewees form the basis for the composition. An assemblage of megaphones, voices and the bodies of twelve female performers swarm out and condense again, effectively creating a temporary mobile speaker system. The Megaphone Choir was performed at press conferences, gatherings, activities and rallies of the Esso-Houses-Initiative.[3] Founded 2011, the initiative has been working to preserve affordable residential and commercial space in St. Pauli. For years, tenants fought tooth and nail to maintain their apartments, shops, and nightclubs. They were backed by the Right to the City (Recht auf Stadt) network in Hamburg[4],

---

2   Amplification: multiplication, potentiation, reinforcement, recruitment, backup, boost, gain, enhancement, strengthener, intensification.
3   Initiative-Esso-Häuser http://www.initiative-esso-haeuser.de.
4   Recht auf Stadt is a Hamburg-based network founded in 2009, which today includes around 40

by the fan base of the football club FC St. Pauli and by neighborhood initiatives like SOS St. Pauli[5]. The initiative continues to accompany the new planning and construction measures on the areal.

My gas station is gone

My gas station

46 years of my life disappeared

46 years I've been living in St. Pauli

46 years are now somehow

OK I guess the last part of St. Pauli

Just died with the Esso Houses

My gas station is gone

Transition accomplished

Fig. 2: Sylvi Kretzschmar/Megaphone Choir Speech Performance during the demolition of the Esso Hauses, Photo by Paul C.P. Krenkler.

initiatives and alliances that stand for affordable housing, non-commercial spaces, socialization of property, a new democratic urban planning, and the preservation of public greens; for the right to the city for all inhabitants – with or without papers.' www.rechtaufstadt.net/category/english/.

5   Initiative SOS St. Pauli: www.sos-stpauli.de.

We amplify a form of speech that was intended neither for public announcement nor political statement. Statements, thoughts, and fantasies of people who would never talk in front of a crowd are developed into dialogue, in the situation of an interview – face to face. Even though some of the interviewees were politically active in Hamburg's the Right to the City movement, or in the Esso Houses Initiative, they only spoke for themselves in the interviews. Often, the speakers can be heard stumbling over their words. It makes a powerful and unusual form of political message by eliminating the eloquent, impassioned speech that we are accustomed to hearing during political agitation.

Megaphones combine a microphone, amplifier and loudspeaker into a mobile and portable PA system. Megaphones' sound has a very directional quality to it. Every movement alters the speech's direction and tone. The transformation of megaphones into a choir requires, but also allows, spatial speech choreography. The megaphones we used also have a recording and playback feature, which we utilized musically for rhythms and loops. In this way, something like to a concrete poetry of political resistance unfolds: a fusion of political speech with modernist and autonomous sound art's creative means.

On the afternoon of 12 February 2014, two giant steel dinosaurs bite into the reinforced concrete and swallow the petrol station that gave the complex of houses its name. All afternoon, I stand at the construction fence, staring at the feasting animals. I begin conversations with former residents, neighbors, passers-by, and activists who are slowly forming a crowd at the gate. The start of the destruction appeared to be the end of a large and diverse protest campaign that had brought the Esso Houses, gentrification processes in St. Pauli, and the resistance of local residents, artists, and activists to the attention of people all over Hamburg. What will be lost when the Esso Houses vanish? That afternoon's interview responses are not aggressive, nor they provide in-depth analysis of the gentrification issue. They are repetitive, they speak to a heart, they are deeply sad, sometimes sentimental and whining; they have their justifications. Interviewees become irritated, they hit a nerve, they build their own poetry.

What happens if you amplify the loss? What happens if you amplify powerlessness? If you DEMONSTRATE the failure, the futility, of a political protest? As a mourning machine, the Megaphone Choir tries to differentiate between grief and resignation in this situation. It is an attempt to deal with the failure of a political initiative collectively, instead of being separated and isolated by its failure.

They'll just turn it all into

Disney World

They sold out

This district completely

Here are shops which

Could be everywhere

In Cologne, in um, well

Honolulu, in Shanghai

Well sorry, um

Sounds embittered

Sounds like always the same but

At the corner

There used to be a hotel

There used to be a small bar...

Fig. 3, 4, 5: Sylvi Kretzschmar/Megaphone Choir: Performance Esso Häuser Echo 2014 Photos by Margit Czenki.

The Spielbudenplatz in front of the construction fence at the Reeperbahn, which was privatized several years ago, became a public venue for the two days of *Megaphone Choir* performances *Esso Häuser Echo*. According to the former residents of the houses, the performances on 24 and 25 of May 2014 fulfilled a concrete function of a funeral for the buildings: a memorial service on the spot.

Before and after the performances, onlookers, passers-by, and Esso Houses Initiative members gathered to discuss what the new buildings should look like. Irene Bude, Olaf Sobczak, and Steffen Jörg, the filmmakers behind *Buy Buy St. Pauli*[6] (a documentary on the Esso Houses) debuted excerpts from their work in progress. Neighbours created an exhibition featuring images of the former apartments and their tenants on the construction fence. For the first time, the PlanBude presented their concept in front of the fence.

PlanBude has been collecting ideas, wishes, sketches, analyses, fantasies, models, plans, and opinions for a new building complex in the Esso-Houses-Area at Spielbudenplatz since October 2014. One of St. Pauli's independent community meetings sparked the idea of starting a planning process from the bottom up: open brief, open to all neighbors, and before the owner or governmental authorities make any decisions. PlanBude provided a wide range of planning tools in two containers near the construction site, allowing everyone to participate in a D-I-Y planning process. To crack the St. Pauli Code, artists, DJs, architects, community workers, cultural and social scientists devised novel approaches such as Lego and plasticine models, photo studies and soundwalks, doorstep interviews, questionnaires for all residences in the neighborhood, and more. PlanBude's work was made possible by protests on the streets. Negotiations with the district mayor, the building department, and the owners worked well. PlanBude is now a much-discussed democratic urban planning model. Its research served as the basis for the architectural competition for the building developments. The Winner NL-Architects (Amsterdam) and BeL-Architects (Cologne) translated the PlanBude results. They designed public roofs, a public balcony and no shopping chains, the return of the music club legend *Molotow,* space for a FabLab, a community canteen and more. The complex of buildings will mix 40% social housing, 20% coop experimental housing and 40% rentable flats. That means 0% condos (no privately owned apartments).

In retrospect, the political protest does not seem to have been a complete failure. The Esso Houses Echo performances were crucial in mourning the buildings and ensuring that the public debate about the area would not cease with their demolition. The performance referred to a proxy speech format that is dedicated to the missing, as an obituary. Issues that need to be discussed from the perspective of former occupants and neighbors are brought to light in a sort of ghost speech at the site of the actual demolition. Others (missing speakers) were speaking over the megaphone choir as a PA system at all times during the performance. It amplified those whose statements were repeated live.

---

6   *Buy Buy St. Pauli* (dir., Irene Bude/Olaf Sobczak/Steffen Jörg, 2014) watch online: www.buybuy-stpauli. de/film-schauen/.

My name is Slatko

I was the last Mohican

The last entrepreneur

With migrant background

At Spielbudenplatz[7]

References

Bude, Irene/Sobczak, Olaf/Jörg, Steffen (dir.). *Buy Buy St. Pauli,* 2014) downloadable at: www.buybuy-stpauli.de/film-schauen/.

Esso Häuser Requiem (the video documentation of the Performance *Esso Häuser Echo* by Sylvi Kretzschmar/Megaphone Choir), (dir. Svenja Baumgardt, 2014), downloadable at: https://www.youtube.com/watch?v=MTdWj04p-7w.

Sankt Paulis starke Frauen – Der Megafonchor – Esso Häuser Requiem (dir. Rasmus Gerlach, 2019), downloadable at: https://vimeo.com/ondemand/179884/260918313.

Initiative Esso Häuser http://www.initiative-esso-haeuser.de.

PlanBude https://planbude.de/category/english/.

PlanBude. *Can design change society? PlanBude - production of desires* Talk by: PlanBude – Hamburg 2018. downloadable at: https://www.youtube.com/watchtime_continue=10&v=uEuXdJVyxIE.

Right to the City Hamburg. www.rechtaufstadt.net/category/english/.

Participatory Art Based Research https://pab-research.de/pabr/.

---

7   Megaphone Choir *Esso Häuser Echo* 2014.

METROPOLIZ/MAAM: THE PRACTICE AND AESTHETICS OF THE URBAN COMMONS IN ROME, ITALY

ANDREEA S. MICU

On 28 March 2009, an ethnically diverse group of people affiliated with BPN, or Blocchi Precari Metropolitani,[1] a Roman housing activist group, forced the locks of industrial space on Via Prenestina 913, in the working class and immigrant area of Tor Sapienza. The group included Moroccans, Tunisians, Eritreans, Sudanese, Polish, Peruvians, Dominicans, Ukrainians, and Romanians, mostly unemployed or precariously employed first-generation immigrants who could not have access to pay rent or buy property in the post-2008 Roman real estate market. The squat was named Metropoliz. In the following decade, Metropoliz has become not only a home for over three hundred people or the first housing occupation in the city that brought together Roma people with other ethnic groups, but also a burgeoning art space known as MAAM (*Museo dell'Altro e dell'Altrove di Metropoliz*, or the Museum of the Other and the Elsewhere).

Fig. 1: View of Metropoliz/MAAM from Via Prenestina. Photo by Andreea S. Micu

Visiting Metropoliz is a remarkable sensorial experience. The old meat factory sits on the side

---

1   Metropolitan Precarious Blocks, in English.

of the road, rectangular shapes of brick and concrete standing against the sky. Even from a distance, the place exerts a powerful fascination on the passerby. The broken glass on the old factory windows, the decaying walls, the general atmosphere of abandonment of the building contrasts with the colorful murals, graffiti and other art pieces that are visible from the outside. Amongst these, the eye readily catches the letters crowning the roof of the construction that is closest to the road and which form the word FART[2]. The tallest part of the construction is a rectangular tower painted in light blue and white, featuring a human figure, an arrow, and the moon in a sequence of three vertical vignettes. An artifact that looks like a telescope sits on top of this tower. Upon walking closer to the entrance, the gaze pauses on the inscription on one of the tower sides, which in blue letters on a semi-circular white background spells the word 'revolution'. The entry gate is made of metal and painted in bright turquoise and black. Hanging on it, several black metallic mailboxes feature dozens of names hand-painted in white. A quick look at these names reveals the diverse ethnic backgrounds of the squatters.

Since its foundation, Metropoliz/MAAM has constituted a vibrant, constantly evolving space. Its evolution illustrates some of the contradictions inherent in practices of squatting as forms of performing the commons, which balance their revolutionary ethos and the need to compromise within existing urban power arrangements to secure housing stability and prevent eviction. In its first decade, Metropoliz has also become a paradigmatic contemporary example of how an art project can help secure the longevity of a housing squat, albeit not free from precarity, critiques, or internal clashes that come from the divergence of opinions. As it is usually the case with squatted spaces, the evolution of Metropoliz has elicited critiques from both insiders and outsiders about what compromises are acceptable as effective political strategies and what might constitute selling out to the interests of capitalist urban development and the white upper middle class values of the Roman art world. TM M hese discussions, often framed as opposing factions in conflict, reflect a fundamental plurality that characterizes the performative construction of the commons. Often, we hear in both popular and academic discourse that such plurality is a shortcoming of contemporary social movements that cannot seem to agree on a specific agenda, as if disagreement were an obstacle for these movements' political goals. The history of Metropoliz, however, points to the limits of this view of plurality as an obstacle to grassroots organizing. A success story in securing housing for its inhabitants for over a decade, the history of this squat precisely reveals that plurality —of discourses, opinions, and actions—is essential to the practice of the commons as a political project that does not seek to cancel difference. The people who inhabit Metropoliz and who are behind the MAAM project hold in productive tension their incommensurable life experiences and identities, as well as their desire to build a common project based on principles on equity, justice, and freedom.

While recent academic work has examined squatting as form of production of the urban commons, mainly from the perspectives of anthropology and urban studies[3], a performance

---

2   Mauro Cuppone, *FART*, 2013.
3   Margherita Grazioli, 'The 'Right to the City' in the Post-welfare Metropolis: Community Building, Autonomous Infrastructures, and Urban Commons in Rome's Self-organised Housing Squats', Doctoral dissertation, University of Leicester, 2017; Alexander Vasudevan, *The Autonomous City: A History of Urban Squatting*, London: Verso Books, 2017; Stavros Stavrides, *Common Space: The City as Commons*, London: Zed Books, 2016.

studies perspective that understands the construction of the commons as a performative practice has been missing from these accounts. To understand the production of the commons as a performative practice, we have to turn to a contemporary Marxist aesthetic approach through the work of Jean-Luc Nancy[4] and José Esteban Muñoz[5], as well as Judith Butler's theorization of 'plural performativity' that 'puts livable life at the forefront of politics' within contemporary assambleary movements against neoliberalism[6]. From such a perspective, it becomes clear that instances of building the commons cannot manifest as self-contained unities, precisely because their nature is to hold a place for difference, staging the persistent and inevitable inter-dependency of life in common.

The construction of the commons is a performative practice insofar as it is realized in/through the embodied behavior of those who come in common seeking collective forms of a good life. This performative practice is inherently plural and open-ended, but it is also embodied in specific actions and gestures that one can learn, teach, repeat, and pass on. It is a practice of the possible, which happens in grand gestures as much as in small everyday ones sustained through a period of time. This practice shows a disposition to make common forms of life not only for collective survival but also collective thriving amidst the economic collapse[7]. The project of understanding the construction of the commons as a performative practice joins recent works in performance studies that look at the role of performance and cultural production in anti-capitalist struggles[8]. More broadly, understanding the construction of the commons as a performative practice speaks to the interdisciplinary scholarship that examines emerging forms of resistance to neoliberal urban development and contemporary housing crises from various fields such as critical urban studies and social movement studies.

4   Jean-Luc Nancy, 'Communism, the Word', in C. Douzinas and S. Zizek (eds), *The Idea of Communism*, London and New York: Verso, 2010; *Being Singular Plural*, Stanford: Stanford University Press, 2000.
5   José Esteban Muñoz, 'Race, Sex, and the Incommensurate: Gary Fisher with Eve Kosofsky Sedwick', in *Queer Futures: Reconsidering Ethics, Activism, and the Political*. Ashgate, 2013.
6   Judith Butler, *Notes toward a performative theory of assembly*, Cambridge and London: Harvard University Press, 2015, p.18.
7   This essay is part of a larger research project that I started as a doctoral student, and that examines contemporary multi-ethnic social projects that build the urban commons by squatting or occupying urban space in Athens, Madrid, and Rome in the aftermath of the 2008 economic crisis. Both drawing on Marxist theory and challenging it from a feminist and minoritarian perspective, I explore practices of constructing an urban commons that have emerged at the intersection of art and politics. My interest is to understand how urban working classes, immigrants, people of color, the urban poor, that is, those most impoverished by the 2008 economic crisis and subsequent austerity policies, use art and performance to *collectively* imagine alternatives to neoliberal urbanization and put them into practice in the South of Europe. As part of this larger project, I visited Metropoliz frequently during my research trips to Rome between 2014 and 2017, attending events that were open to the public, such as the MAAM's open Saturdays, as well as some other non-public events, such as some Tuesday internal assemblies. I have spoken formally and semi-formally with artists whose work is featured in the museum, some of the squatters, and others housing activists in the BPM network.
8   Stefano Harney and Fred Moten, *The Undercommons: Fugitive Planning & Black Study,* New York: Minor Compositions, 2013; Joshua Chambers-Letson, *After the Party: A Manifesto for Queer of Color Life*, New York: New York University Press, 2018; José Esteban Muñoz, *The Sense of Brown*, Durham: Duke University Press, 2020; Judith Hamera, *Unfinished Business: Michael Jackson, Detroit, and the Figural Economy of American Deindustrialization*, New York: Oxford University Press, 2017.

A Museum in the Urban Periphery

The part of the urban periphery where Metropoliz is located looks very different from the monumental Rome that attracts millions of visitors every year. Instead of the usual hordes of tourists holding cameras and selfie sticks, or the young professionals going to their offices in the city center, when taking a bus or tram to Tor Sapienza one finds an almost exclusively immigrant working-class population. Via Prenestina runs for miles to the outskirts of the city, through poor and working-class neighborhoods, empty lots, and unsightly commercial areas with furniture stores, gas stations, supermarkets, and transportation and auto repair businesses.

Tor Sapienza lived its golden years in the 1960s when the factories in the surrounding areas provided full employment for its entirely Italian working classes. Then, starting with the oil crisis of the 1970s and into the 1980s, factories closed or moved to other locations. These shifts in production made some of the population move out of the neighborhood following the available jobs. Small family-owned businesses started to close, a trend that later continued with the arrival of big supermarkets and shopping areas. While the degradation of the neighborhood was caused by deindustrialization and rising unemployment, many of its old inhabitants blamed the successive migratory waves that were arriving in the 1990s[9]. These waves gradually changed the demographic composition of the neighborhood, but also marked the existing difficulties of integration and coexistence that were being formed alongside race and ethnicity lines. No group encountered more hostility than Roma people from Eastern Europe, which first started arriving as refugees of the wars in former Yugoslavia. The city administration built a camp on Via Salviati to lodge them until permanent housing could be provided, but the camp, which was supposed to be only temporary, has never been dismantled. In addition to the Roma camp on Via Salvati, Tor Sapienza has a centro di acoglienza (asylum center) for immigrants on Via Morandi. Mostly occupied by both Northern and Sub-Saharan Africans—and more recently refugees from the wars in the Middle East—the asylum center is chronically overcrowded and lacking in resources. Racism and xenophobia are pervasive in the area.

Metropoliz emerged in this geographical and social context. From the inside, the squatted factory is as appealing as one would imagine. Beyond the entry gate, a yard opens to the right and an alley that goes straight to the end of the factory limits. Most living units are at that farthest end from the street. A few of them are also on the right side of the yard, by the entry gate. These are small housing constructions painted in white and light blue, with some pots of flowers on the windows and children toys dispersed at the front. Scattered here and there around the yard, one usually finds supermarket carts with discarded materials that some of the squatters collect from the street. Reselling old appliances and electronic debris for the metals they contain is a common source of income for many unemployed Roma people living in Italy. Across from the entry gate, a few steps and a ramp mark the access to the main building of the former industrial complex, where the museum is now located. This large building is a labyrinth of tall and scantily lit galleries around a roofless central space. This space now serves as a patio of sorts, letting the light come in towards the surrounding areas. All the galleries around this central one are made of

9   Adriana, Goni Mazzitelli, 'Il Ruolo di Space Metropoliz in una Pidgin City in Divenire' in Fabrizio Boni and Giorgio de Finis (eds), *Space Metropoliz: L'Era delle Migrazioni Esoplanetarie*, Bordeaux, 2015.

concrete walls and concrete floors. Some walls are still covered with the white tiles one can find in butcher shops, a reminder of the animal slaughtering activities that the factory used to host.

By most accounts, the creation of what is today the MAAM, the Museum of the Other and the Elsewhere happened out of chance. In 2009, during the same period that BPM was squatting the former Fiorucci meat factory to found Metropoliz, in a different side of the city, a number of architects, artists, and activists tied to the Stalker collective[10] were developing participatory tools and actions to create what they called 'collective imaginaries' for places of the urban periphery. Anthropologist Giorgio de Finis and filmmaker Fabrizio Boni were two of the participants in the Stalker activities. In 2011, they were looking for a place in the city periphery that could host their idea of creating a rocket 'to migrate to the moon' and shoot a documentary of the process. The moon was the metaphor for a public space that belonged to all humankind and that could not be privatized and endangered by capital operations. De Finis and Boni arrived at Metropoliz with this proposition, and after gathering the interest of the squatters, they started shooting Space Metropoliz. During those months, a few visual artists intervened on the old factory walls making pieces that reflected on the topic of human migration to the moon as a utopian egalitarian project. Those first pieces were the beginning of what would later become the MAAM. After finishing shooting the documentary, Giorgio de Finis continued his engagement with Metropoliz, opening the space to the participation of local artists who donated their work in support of the squat and its housing rights struggle. In April of 2012, and from the donated pieces of dozens of artists, the MAAM was born as a 'political museum', in De Finis's words.

Fig. 2: View of the roofless central patio of the MAAM, and some of its art pieces and murals. Photo by Andreea S. Micu.

10   Stalker is a collective of architects, activists, artists, and researchers connected to the Roma Tre University founded in 2002. They work experimentally and engaging in actions to create self-organized urban spaces.

(In)Definitions of Space

Unlike most conventional galleries or museums, the inside of the MAAM is not a white neutral box meant to direct the viewer's focus to the individual art pieces. Different artistic works overlap on the walls with stains of smoke and humidity, chipped materials, and previous graffiti that had been made during the time between the closing of the meat factory and the arrival of the squatters. There is a casual aspect of messiness in the space, in the way the artworks stand on the walls, competing for the viewer's eye. And yet, it is this very messiness that constitutes the MAAM's unique aesthetic appeal, as if the whole space were one single collective art piece, a huge mosaic made by individual artworks.

This messiness of the space is replicated in the differences of opinions and disagreements about the role of the art space in the squat. For De Finis, the existence of the museum inside the squat fulfills a social function, guaranteeing the protection of the site against demolition. The land on which Metropoliz sits was purchased a few years ago by the Salini Impregilo Group, one of the biggest construction companies in Italy. Like other former industrial spaces, this piece of land is appealing to real estate speculators. In Rome, industrial land is priced considerably lower than building land, and therefore, just by buying an industrial property and getting public administrations to change its designation to land suitable for building, real estate developers make a significant profit. The Salini Impregilo Group intends to eventually demolish the factory and use the land to build apartment buildings, but the BPM occupation of the space trumped the project. To some extent, the extensive activist network of BPM, which has several large housing squats all over Rome, helps guaranteeing the permanence of Metropoliz. In the past, BPM has proved itself able to galvanize thousands of supporters to demonstrate against evictions or make camp in a given property to obstruct police access. But, according to De Finis, the consolidation of MAAM as a contemporary art space is at least as important in the permanence of the squat as the housing activists's labor.

Fig. 3: Hall to the left of the main entry, MAAM. Photo by Andreea S. Micu.

Certainly, the existence of the art space has been taken into consideration by the construction business. Pietro Salini, the business mogul and CEO of Salini Impregilo, has summoned Giorgio De Finis to his luxurious office located in the city center next to the Fontana di Trevi. De Finis recalls the details of that meeting:

> *'What are we going to do with Metropoliz?,' he asked me. And I looked back at him and I said: 'What are we going to do with it?' And then he asked me whether I would accept keeping the museum intact and all the art in it and my place as a curator, but just evict the people wholive here. And I said, without the people, the museum means nothing.Precisely what makes it different is that it is a museum, and a homefor those who don't have a home. And it's both things. Otherwise, it would be just an old factory full of painted walls. That's what I told him. Because, you know, these artworks have emerged as a response to the situation of the squat's inhabitants, and they don't make sense out of this context. And you know, the sad thing is that if you evict three hundred people, nobody cares, but if you destroy an art space, then everyone cares because you look like an animal that doesn't appreciate culture. I mean it is sad, but it is also our biggest advantage. In the end, we didn't reach any agreement, but we ended up in polite terms.*

In opposition to Salini's idea to evict the squatters and keep the MAAM, De Finis proposed an alternative: the Salini group would invest some money into the living spaces and donate them to their current inhabitants as a sort of *altruistic project* for the city. According to De Finis, this would incidentally be a great benefit for the Salini Impregilo corporate image. He tells me that he assured Salini that they could do 'something better than the Venezia Biennale'. Of course, he admits that for Salini letting the squatters stay could be a delicate matter, because it would create a precedent in a city like Rome with so many occupations and squats, strengthening the model according to which a squat that has artistic and cultural value is more likely to avoid eviction.

Salini's plans for the old factory have never aligned with Giorgio's proposal, but in the last few years ho has perhaps become more aware of how the art inside MAAM might boost the value of his real estate. Upon filling a lawsuit against the squatters and going to trial in 2016, Salini's attorney stated that their client intended to keep the art after the eviction of the squatters. And yet, much to Salini's chagrin, and due to the indecision of the Italian state to enforce the eviction and deal with the unpleasant consequences that might come from it, the museum has so far succeeded in protecting the squat. Eventually, Salini also sued the Italian Ministry of Interior for failing to enforce the eviction and allegedly damaging his private property rights. In July 2018, a Roman civil court deemed that the Ministry owed 28 million Euros to Salini Impregilo that are yet to be paid at the time of this writing.

Court decisions aside, Giorgio's need to at least sit and talk with a construction conglomerate like Salini Impregilo points to one of the limits of the squat as a radical practice of sovereignty. This limit is marked by the constraints imposed by existing land property laws in Rome and the power and political influence of Salini Impregilo. De

Finis's initial intention to convince Pietro Salini to donate the space to the squatters as a gesture of corporate charity is arguably a strategy of survival. It is also evidence that his investment in the space is directly related to the existence of the art space. Regardless of whether De Finis is aware of the irony of his reference to the Venezia Biennale and the increasing criticism that the art biennale model has received in the last few years as examples of art spaces that are co-opted by capital and funded by corporations[11], what seems clear is that thinking of the MAAM in such a term implies that he is fully aware that the MAAM is a space with broad appeal in the art world. As part of this appeal, political activism and squatting are selling points in the process, constituting experiences that certain audiences are eager to consume.

For De Finis, taking steps towards the institutionalization of the art space is a positive and desirable strategy because it attracts public support for the MAAM and, indirectly, for the squatters. He proudly talked to me about national and local political figures and art celebrities that have visited the MAAM and been impressed by it, even 'a former Minister of Culture'. Indeed, it seems that the story of a squatted factory full of artworks is appealing, as in the time of my last visit to the site in January 2016, De Finis had secured a collaboration between the MAAM and the Rivoli Museum in Torino, one of the most important contemporary art institutions in Italy. And in 2018, De Finis was made director of the contemporary art museum Macro by the Roman public administration of arts and culture, a position that he combined with his role at MAAM. De Finis is not naive about the power that Salini holds over the squat and the art project inside, but he seems convinced that the more consolidated and institutionally recognized the art project becomes, the easier it will be for the squat to survive. To his credit, the first decade of Metropoliz existence seems to confirm his view.

For other people inside the squat and in the broader Roman housing movement of BPM, however, trying to convince Salini to make Metropoliz into a project of corporate responsibility always felt like an unacceptable capitulation to the interests of power and capital. For them, De Finis new position at Macro only proves that his priority was always his curatorial career over the interest of the squat. The very consolidation of MAAM as an institutional art space is in this view a matter of controversy. From this perspective, the current path that the MAAM is taking towards institutionalization is a betrayal to the radical potential of squatting and reclaiming urban space from the interests of capital. One of the most critical voices, who was an active part of the squat at the beginning but withdrew later for ideological disagreements over the course that the art space was taking, went as far as asserting that the MAAM is capitulating to gentrification. For this interlocutor, the MAAM's current selection of artists is also a problem:

---

11   Yahya M. Madra, 'From Imperialism to Transnational Capitalism: The Venice Biennial as a
      "Transnational Conjuncture"' *Rethinking Marxism* 18.4 (2006): 525; Panos Kompatsiaris, *The Politics of
      Contemporary Art Biennials: Spectacles of Critique, Theory and Art*, New York and London: Routledge,
      2017; Jeannine Tang, 'Of Biennials and Biennialists: Venice, Documenta, Münster', *Theory, Culture &
      Society* 24.7-8 (2007).

It used to be that anyone who wanted to collaborate and support the squat could go and do their work there. Now, because they are more and more aware of their self-importance as a museum, the artists who already have a name are given more space, and their work is placed in more visible places, and the artists who nobody knows are relegated to some corner. And the place is increasingly receiving a bourgeois audience who visits on Saturdays, take a walk, have some exotic food, and take a picture of a Roma child like they would do with a monkey at the zoo, and then they can go back home feeling good about having taken a tour to the city's periphery to experience how marginalized people live for a few hours.

This interlocutor was referring to the urban middle class, who increasingly composes the audience that visits the MAAM on Saturdays to enjoy the radical aesthetics of anarchism and revolution, without any of its material risks. During my visits to the museum's open Saturdays, I saw the presence of these groups. These visitors seemed to be a mix of university students, art practitioners, hipsters, middle-age social democrats, and middle-class families who consume contemporary art for leisure on the weekends. And although I never saw anyone taking a picture of a Roma child, I did see a number of art-savvy—one could say, following my interlocutor, bourgeois—spectators from more affluent parts of Rome behaving exactly like they would do at a conventional museum or art gallery space. Some of these people walked around in awe, took pictures, and stopped in front of the pieces with contemplative attitude, commenting about the quality of the work, as if the quality of the work could be divorced from the material conditions in which it was produced. During these Saturdays, the interactions between the squatters and the visitors were limited. The people at Metropoliz provided the services, such as cleaning the space, welcoming visitors by the door, or preparing the food, much like working-class people of color would do in a different, more conventional cultural institution.

For others inside the squat, opening the space to certain middle-class audiences was desirable. Arguably, these audiences marked Metropoliz as a space that was different from other social squats in Rome, perceived to be much more *unruly* and marginal. This was Mustafa's position, as he explained:

We don't want to be like other social centers that organize concerts and parties and make a lot of money out of it. We don't do it for the money. If we did, we could get a lot more, but in those places there's loud music, and people get drunk and do drugs, and that's not the image we want to have in the neighborhood. We have an art space, so people can come here with children. We get a lot of families that come to visit.

Mustafa usually stands by the entry gate on Saturdays. He is in his forties and comes from a rural area in the North of Morocco, although he spent a few years in Casablanca before coming to Italy. He is in charge of informing visitors about the donation policy. Although it is not required to pay in order to visit the MAAM, he advises people to donate a minimum of five euros. He has a recurrent joke that I have seen him perform many times. It is usually some variation of 'If you want to leave a hundred, or a thousand euros, it's up to you. I'm not gonna say no!' Visitors laugh and leave some coins, or a five, ten, and in very rare cases, a twenty-euro bill in a wooden box that sits on a squeaky portable table in front of Mustafa. He has a natural talent for talking to all kinds of visitors and a friendly demeanor that makes people want to talk to him. During the times he is standing by the door waiting for visitors, one can stop by and chat with him about almost everything involving current international and Italian politics.

For Mustafa, attracting middle-class families to visit the MAAM is actually a strategy to consolidate Metropoliz's permanence in the neighborhood. Others in the squat share his view, aware of the need to be perceived as conducting a respectable social and cultural struggle, and not just squatting illegally. Taking into account that Metropoliz hosts a high number of Roma people, always affected by stereotypes of incivility, dirtiness, and unruly behavior, this desire is understandable. Not everyone has the privilege of refusing respectability as a bourgeois value while they are constantly subjected to marginalizing stereotypes and depicted as savage. From this position, the art inside the squat is a legitimizing instrument for their housing struggle.

If Salini were to comply with De Finis's proposition of giving Metropoliz to the squatters, this would certainly solve their immediate need to have a house. But would it represent a sustainable model for affordable housing in Rome? Would it stop any of the macroeconomic urban development processes due to which these people did not have a house in the first place? Would it contest the involvement and responsibility of companies like Salini Impregilo and their speculative practices in the existing housing problems of the city? Moreover, would finally having a house with all due legal protections deter some of these squatters from struggling alongside the broader housing social movements in Rome? This last question is an essential part of the paradox of struggling for housing within an existing legal system that privileges models of private over communal property. In this context, fighting for the legal status of existing squats is for squatters both necessary in the short term as a solution to their housing problem, and in the long term, a limit to their own movement of production of the commons. Arguably, achieving legal recognition for squatted properties cannot stop the macroeconomic processes of capitalist urban development that constantly create homeless people. And while many squatters might be aware of this, they are also under the predicament of securing a living space; a need that cannot be endlessly postponed if and when there is a chance of achieving some form of legal recognition to stay.

Chatting with the squatters, and people involved in BPM and the broader housing struggles of the city, one is left with the impression that there are almost as many versions about what Metropoliz/MAAM is and what is supposed to do as a political, artistic, and social space as there are people involved in it. But is this variety of opinions an obstacle for the project of constructing the commons that the squatters are involved in? Or rather, how are we to understand the commons in order to make place for plurality? The final section of this essay proposes that the performative nature of building the commons happens precisely in the dialogical space of irreducible differences of those coming together.

Fig. 4: View of one of the open patios between buildings, MAAM. Photo by Andreea S. Micu.

A Theoretical Approach to the Commons

Building the commons is a performative practice of production of material, tangible conditions sustaining the life of the human body that in turn enable the production of an anti-capitalist subjectivity. Three interrelated dimensions characterize the production of the commons: materiality, subjectivity, and practice.

As material production, the commons are concerned with human needs that depend on infrastructure and resource distribution, such as food, water, shelter, healthcare, education, labor, etc. In our historical present, there is no production of the commons that is not, at the same time, a struggle to protect or recuperate the material dimensions of life from the processes of neoliberal capitalism. It is to this material dimension of human life that Jean-Luc Nancy refers when asserting that "the truth of the common is property"[12]. And it is with this material dimension of human life that Metropoliz's squatters concern themselves. Shelter, or the right to proper dignified housing conditions is their fundamental demand. But housing is never isolated from other material conditions that hinder or facilitate one's access to housing, such as labor. And labor, in turn, is inseparable from the conditions in which the laboring body operates and that determine its ability to perform such labor, such as housing, healthcare, education, etc. The multiple material needs that sustain the human body are inextricably linked, and therefore, there is a certain totalizing—

---

12   Nancy, 'Communism, the Word', p.149

one might say holistic–disposition of the commons to address all of them, such that even when a common project focuses on a particular need, such as housing, the others are inherently present.

Nancy's definition of property exceeds the material, taking us into the second dimension of the commons stated above: subjectivity. Nancy asserts that:

> Communism has more than, and something other than, a political meaning. It says something about property. Property is not only the possession of goods. It is precisely beyond (and/or behind) any juridical assumption of a possession. It is what makes any kind of possession properly the possession of a subject, that is, properly its expression. Property is not my possession; it is me.[13]

For Nancy, property and subjectivity are inseparable, such that the material conditions of life–what we have and how we have it–determine the realization of who we are. In this, Nancy follows Marx's axiom that mankind produces its history, and in doing so, produces itself.

Drawing on Nancy, performance studies scholar José Esteban Muñoz interrogates the commons from the perspective of minoritarian subjectivities. Both Nancy and Muñoz are concerned with thinking about the possibilities of togetherness while refusing an understanding of community as unity. Unity or equivalence is not only ontologically impossible but even as an ideal is ultimately mobilized on erasing difference. One of the essential problem for the commons, for Muñoz, is how to think togetherness across radical differences that in the moment of singulars coming together are already embedded in a power structure. Muñoz contends that a queer of color subjectivity can offer us a glimpse into 'a commons of the incommensurate that signals something that goes beyond a politics of equivalence'[14] He goes on to argue that communism would be antithetical to colonialism, both structural and internalized:

> Ideologies that enable empire are shored up by a reification of the individual sovereign subject who can think of itself as differentiated from a larger sense of the commons. Thinking of the self as purely singular enables a mode of imagining the self as not imbricated in a larger circuit of belonging, what I call an actual sense of the world where we grasp the plurality of the senses, which is not one's own senses but instead the multiple senses of plural singularities.[15]

Muñoz's choice of the words 'imagining the self' is particularly pertinent insofar as it points to a construct in which the relationality of existence is ignored in favor of a colonial imaginary according to which individuals are isolated monads. And more importantly, 'imagining the self' implies a fundamental relationship between imagination and practice, between what one would like their life to be and what one does to make it actual. Following Muñoz, we see how the individual isolated from the common is a colonial project, and more importantly, that being in common implies a togetherness of what is incommensurable across singularities.

---

13   Nancy, 'Communism, the Word', p.148
14   Muñoz, 'Race, Sex, and the Incommensurate', p.112
15   Muñoz, 'Race, Sex, and the Incommensurate', p.113

Following Nancy, and Muñoz, we might ask: if capitalism enables the illusion of self-contained subjectivities, what kind of forms of the self might arise from communal forms of property? Who is the *proper* self of the commons? For Muñoz, this mode of being is available to us as sense, always necessarily entangled with the senses of plural singularities. The singular that Muñoz talks about is bound to sense its multiplicity, its relational existence as simultaneously one and many. This is a singular self who knows of its unbreakable interdependence with others. This is the *proper* self of the commons.

It is in the essential link between the commons as material property, practice, and the formation of subjectivity where the radical potential of squatting lies, in general, and where the radical potential of Metropoliz/MAAM lies as a political project, specifically. Beyond the physical immediacy of occupying urban space and claiming it for purposes other than capital accumulation, beyond the urgency of providing homes for the homeless, squatting is a performative practice of the commons, one that enables forms of sociality guided by an egalitarian ethos. In these performative practices, material conditions and subjectivities constitute each other in ways not easily subsumed to the dictates of the capitalist (neo) colonial ordering of the world. The commons are a performative practice because they are an instance of world-making, or tangible material transformation towards certain living conditions perceived to be more just than the existing ones.

The production of the commons in the sense that Nancy and Muñoz posit has to account for incommensurable difference in ways that neither subsumes it to uniform collectivity nor uses it as ultimate evidence of why different singularities cannot coalesce in common political projects. This would be a construction of the commons that foregrounds the relational, that makes place for that which we cannot do away with in order to step into a neutral public space, open to the multiple differences of bodies that need different things in order to carry a livable life. But it so happens that one does not get to choose whom to construct the commons with, because one ultimately cannot do so. And yet the question remains of how to still nurture forms of commoning in which we enter from radically different perspectives and personal experiences, something that Judith Butler has pithily expressed as follows:

The people you find in the street or off the street or in prison or on the periphery, on the path that still is no street, or in whatever basement that houses the coalition that is possible at the moment are not precisely the ones you choose. I mean, for the most part, when we arrive, we do not know who else is arriving, which means that we accept a kind of unchosen dimension to our solidarity with others. Perhaps we could say that the body is always exposed to people and impressions it does not have a say about, does not get to predict or fully control, and that these conditions of social embodiment are those we have not fully brokered. I want to suggest that solidarity emerges from this rather than from deliberate agreements we enter knowingly.[16] For Butler, what Muñoz defines as incommensurable is precisely the terrain of solidarity.

When people come together in a project like Metropoliz/MAAM they do not simply get to start from zero, as if their bodies and identities were not already constituted by power and

16   Butler, *Notes toward a performative theory of assembly*, p.152

the very forms of domination that they try to leave behind. The potentiality of the commons resides in the simultaneous coming together in essential difference that is marked by existing behaviors, practices, and discourses, but it also resides in coming together to undo neoliberal relationality and to make anew what *being together* means. The project of being in common is thus the open-ended process that starts from this paradox.

The practice of commoning happening at Metropoliz/MAAM, the irreducible plurality of experiences, and the encounter of radical differences among the squatters is the very fabric from which the commons emerge. This approach to understanding the commons emphasizes process over results, suggesting that practices of commoning such as the one happening at Metropoliz/MAAM have value beyond their immediate results. A practice is constituted by that which we make and remake, and that in turn makes us. Practices are carried through a period of time, learned and taught to each other. They can also be transformed to adapt to different circumstances and needs. What we practice is learned and remembered by the body, shaping who we are through the slow ongoing sedimentation of repetition.

Practices of commoning are stored and remembered by bodies and can therefore travel beyond the immediate project of squatting a particular property. Even if Metropoliz/MAAM is demolished and transformed into apartment buildings by Salini Impregilo and its current dwellers evicted, the political potential of their project of the commons will not be extinguished. For Metropoliz/MAAM dwellers, the real project of building urban commons is not—or not only—the squatting of a former industrial space, but the embodied, learning process that they have experienced together. This is a practice that they could take somewhere else, some other time. It is a practice that can live on despite the eventual cranes and excavators that might transform their commune into apartment buildings for the sake of capital accumulation.

References

Butler, Judith. *Notes toward a performative theory of assembly*, Cambridge and London: Harvard University Press, 2015.

Chambers-Letson, Joshua, *After the Party: A Manifesto for Queer of Color Life*, New York: New York University Press, 2018

Goni Mazzitelli, Adriana. 'Il Ruolo di Space Metropoliz in una Pidgin City in Divenire' in Fabrizio Boni and Giorgio de Finis (eds), *Space Metropoliz: L'Era delle Migrazioni Esoplanetarie*, Bordeaux, 2015, p. 100.

Grazioli, Margherita. 'The 'Right to the City' in the Post-welfare Metropolis: Community Building, Autonomous Infrastructures, and Urban Commons in Rome's Self-organised Housing Squats'. Doctoral dissertation, University of Leicester, 2017.

Hamera, Judith. *Unfinished Business: Michael Jackson, Detroit, and the Figural Economy of American Deindustrialization*, New York: Oxford University Press, 2017.

Harney, Stefano and Fred Moten. *The Undercommons: Fugitive Planning & Black Study,* New York: Minor Compositions, 2013.

Kompatsiaris, Panos. *The Politics of Contemporary Art Biennials: Spectacles of Critique, Theory and Art*, New York and London: Routledge, 2017.

Luisetti et all. (eds) *The Anomie of the Earth: Philosophy, Politics, and Autonomy in Europe and the Americas,* Durham and London: Duke University Press, 2015.

Madra, Yahya M. 'From Imperialism to Transnational Capitalism: The Venice Biennial as a 'Transnational Conjuncture'' *Rethinking Marxism* 18.4 (2006): 525-537.

Muñoz, José Esteban. 'Race, Sex, and the Incommensurate: Gary Fisher with Eve Kosofsky Sedwick', in *Queer Futures: Reconsidering Ethics, Activism, and the Political.* Ashgate, 2013: 103-115.

Muñoz, José Esteban. *The Sense of Brown*, Durham: Duke University Press, 2020

Nancy, Jean-Luc. "Communism, the Word," in C. Douzinas and S. Zizek (eds), *The Idea of Communism*, London and New York: Verso, 2010.

—————. *Being Singular Plural*, Stanford: Stanford University Press, 2000

Stavrides, Stavros. *Common Space: The City as Commons,* London: Zed Books, 2016.

Tang, Jeannine. 'Of Biennials and Biennialists: Venice, Documenta, Münster', *Theory, Culture & Society* 24.7-8 (2007): 247-260.

Vasudevan, Alexander. *The Autonomous City: A History of Urban Squatting*, London: Verso Books, 2017.

THIS BUILDING TALKS TRULY

FILIP JOVANOVSKI, IVANA VASEVA & KRISTINA LELOVAC

Fig. 1: Railway Residential Building. Photo by Zoran Shekerov

Our contribution in this book is the script for the performance that was presented for the first time at the Prague Quadrennial of Performance Design and Space 2019 in Prague, as a part of the installation *This Building Talks Truly*. The main scene of the performative installation is the Railway Residential Building in Skopje. The building is an active participant together with the actress Kristina Lelovac and the audience that are summoned to become potential tenants or members of the household council. The scene is built from a mobile stage elements and attached video projections, sound speakers and written text boards. The performance narative speaks about the problems that the residents of this building have been facing. The stories have been collected during the research phase that has included the survey and an attempt to test the possibilities for the common resolution of existing issues with the help of the forum theatre technique.

The Railway Residential Building in Skopje raises important questions about privatization and commodification of commons in Yugoslav, anti-Yugoslav and post-Yugoslav times. Although these once societal-apartments have been already privatized, in the comprehensive, yet silent privatization processes within the frame of the post-socialist transition that has meant the restoration of capitalism, Yugoslav architectural concept for collective housing that this building embodies still provides a space and an opportunity for reasoning about different models for collective living and collective management of common spaces.

Fig. 2: If buildings could talk[1] - First performance within the festival Young Open Theater –Skopje (2017). Photo by Zoran Shekerov.

Act 1: History as Stage

The red curtain is down. A microphone on a stall is placed centrally in front of the curtain. The Performer dressed in dark blue workers' topcoat walks in front of the red curtain and rises a card board, on which 'Act 1. History as Stage' is written.

THE PERFORMER: Act 1! History as Stage! [*With sharp movement, she points the cardboard to the audience in the front, on the left and on the right, for 5 seconds to each side. With sharp movement, she puts down the cardboard. She goes to the microphone and greets the audience*] Comrades, today on *actual day and month of performance*, 1949, we are gathered here in Skopje, the capital of People's Republic of Macedonia, a part of the People's Federal Republic of Yugoslavia, to celebrate the opening of the Railway Residential Building. It is honor to have today with us the building's originator Michail Dvornikov [*She points to someone in the audience. Short pause*], and our comrade from Belgrade, the architect, sculptor and theorist Bogdan Bogdanović. [*She points to someone in the audience. Short pause*] The Railway Residential Building is intended for the employees of the State Railway Directorate in Skopje, a part of the Federal Yugoslav Railway. The building is an example of an object with a combined program. Apart from the residential units, there are also semi-public, common

1   *If buildings could talk*, Research based performative art project by Filip Jovanovski, curated by
      Ivana Vaseva 01 October 2017 - First performance within the festival *Young Open Theater* in Skopje,
      Participants and collaborators: Kristina Lelovac, Sanja Arsovska, Jasmina Vasileva, Dolores Popovikj,
      Ivana Pavlakovikj, Tamara Ristoska, Blagoja Veselinov, Vlado

are as intended to be shared by the tenants — the central courtyard, the open terraces, the common laundries on the top floors, and the cinema. [*Two stage workers raise the curtain. The Performer points to the concrete-like front wall of the installation. Pause*] Comrades, this is the Railway Residential Building! [*Pause. She moves the microphone stall to the side and takes of the topcoat*] This is not the actual opening of the Railway Residential Building. This is not a theatre performance about that opening. This is not theatre. Almost not at all. But, [*she points at the installation*] this is the Railway Residential Building. [*pause*]

The Railway Residential Building in Skopje was built at the same time when Le Corbusier was building the Marseilles Bloc. [*She rises a cardboard with a picture of Le Corbusier. With sharp movement, she points to the audience in the front, on the left and on the right, for 5 seconds to each side. With sharp movement, she puts down the cardboard*] The Marseilles Bloc is inscribed as UNESCO World Heritage Site. Le Corbusier imagined and created buildings to be machines for living. One building is one city! [*pause*] At the end of the 1940s Yugoslavia separated from the Soviet Union. Tito said NO to Stalin. [*With sharp movement, she flips the cardboard and rises it. On this side of the cardboard is a picture of Tito, Stalin and a big NO. The installation starts to slowly move back. With sharp movement, she points it to the audience in the front, on the left and on the right, for 5 seconds to each side. With sharp movement, she puts down the* cardboard] The Communist Party organized a gathering in the cinema in the building and announced to its members, for the first time in than People's Republic of Macedonia, Yugoslavia's separation from the Soviets. The event was filmed by Trajče Popov. I suggest you to remember this information because you will need it afterwards. After the separation from the Soviets, in Yugoslavia an entirely new approach to art was introduced- favoring the abstract art of the Western world as a counterpoint to the ongoing Stalinist socialist realism. One of the most representative examples of the Yugoslav socialist modernism are the monuments devoted to the Yugoslavian antifascist movement. The most prominent author of this paradigm is the architect, sculptor and theorist Bogdan Bogdanović. [*She throws the cardboard on the floor and takes another one from the side*] This is a story of three children born and raised in the Railway Residential Building. They are not children anymore. Ivan is a pensioner, Bajo used to be a basketball coach and Tome is an architect-technician. As children of the employees of the Railway, they inherited the apartments from their parents. However, in the 1990s they had to buy them from the state again. This is a story about their memories of the building. [*Pause*] I also had a friend who lived briefly in this building. As many others did in the last few years, he left Skopje for good. In the first few months after he left, he used to send me letters. Not e-mails, but letters on paper, you know? Old school. In one of them, he wrote to me... [*With sharp movement, she rises a cardboard on which the following is written: "I will have spent my life trying to understand the function of remembering, which is not the opposite of forgetting, but rather its lining. We do not remember; we rewrite memory much as history is rewritten. How can one remember thirst?*[2]*" With sharp movement, she points the cardboard to the audience in the front, on the left and on the right, for 15 seconds to each side. With sharp movement, she flips the cardboard. On the other side 'ACT 2. Entrances' is written*] Act 2! Entrances! [*She throws the cardboard on the floor and takes another one from the side*]

---

2   Sans Soleil, Chris Marker, 1983, https://www.markertext.com/sans_soleil.htm.

Fig. 3: Ivan Dzijanovski, Slobodan Kocevic and Tome Karevski, residents at Railway Residential Building. Photo by Zoran Shekerov.

Act 2: Entrances

THE PERFORMER: [*Welcomes the audience to get inside the installation, in front of the concrete-like wall*] Come, you can get inside. It's a bit tight but how else can it be, we are a crowd in an entrance of a building. This is the Railway Residential Building's entrance no.1, one of the ten entrances of this building. This is the apartment no. 1 on the ground floor. In the 1950s it was divided into two parts and shred by two families. In one part lived the family of Antonio Kuzmanovski, who moved right after the building was open. During the 1950s, they shared the apartment with the family of Schulz. We asked the tenants, but nobody remembers his given name. What they remember is him being extremely high... like... [*She looks to the highest person in the audience*] Can you please come for a second? Can you stand on the construction? And now, can you go up on your toes? If you need balance you can hold my hand. [*She puts her other hand above the men's/women's head*] Yes, this is how Shultz was tall. Thank you. [*The man/woman goes back among the audience*] He received the nickname after a very popular acrobat from the circus that was occasionally visiting Skopje in the 1950s. Even today, the tenants, remember that all members of the Schulz family were kind of corpulent. Schulz's first son was the first child born in the Railway Residential Building. [*Pause*] In 1992, in this same apartment, for just one night, stayed the architect, sculptor and theorist Bogdan Bogdanović. He is the author of the prominent monument *Stone Flower* [*She rises the cardboard with a picture of the monument. With sharp movement, she points to the audience in the front, on the left and on the right, for 5 seconds to each side. With sharp movement, she puts down the cardboard*] dedicated to the victims of the concentration camp in Jasenovac, in today's Croatia. He is also the author of a memorial park *Mound of the Unbeaten* [*She flips and rises the cardboard. On this side there is a picture of the monumental park. With sharp movement, she points it to the audience in the front, on the left and on the right, for 5 seconds to each side. With sharp movement, she puts down the cardboard*] in the town Prilep, in today's North Macedonia. Bogdanović was antifascist, he was antinationalist and he supported the anti-war movement in Yugoslavia in 1990s. Being oppressed by the Milošević regime, he had to leave the country and move to Vienna. On his way to Vienna, for just one night, Bogdanović stayed in this apartment. Before leaving, he wrote a note on one of the walls. [*She takes out a square part of the wall. Under it a cardboard with a tick packing paper appears, and on it the following is handwritten: 'If I remained silent, today I'd live peacefully yet ashamed. But it was impossible to be silent'. Pause*] Bogdanović's writings on the wall was painted over [*with sharp movement, the Performer removes that piece of paper and another blank piece of paper appears underneath*] in the early 2000s when the apartment was bought by Zan Stefanovski. Zan Stefanovski is a former Macedonian architect who is known as the author of the 66-meter high tower called the Millennial Cross. [*With sharp movement, she draws the monument on the paper with a black tick marker*] Yes, it looks exactly like this. It was erected on the top of Vodno [*as a child drawing, she draws a mountain underneath the tower*] — a mountain in the heart of Skopje. Since then, we have a huge cross on the top of the city so it looks just like a grave. It was erected at the end of the 1990s to announce the coming of the new millennium. Unfortunately, back than we were not aware that it was also an announcement of something else [*with sharp movement, she writes over the drawing 'Skopje 2014'*] — the controversial project *Skopje 2014*. The project was initiated by the previous Macedonian Government. The ruling party was right-wing, conservative and nationalist. They said that they are initiating this project to renovate the city center to look older, to give it a more classical appeal. They spent. . . [*On the paper she writes 600. Pause. She writes three zeros. Pause. She writes a dot after the third zero. Pause. Slowly and looking at the audience, she writes 3 more zeros. Pause. Again, slowly and looking at the audience, she writes the sign of Euro*] . . . on monuments, baroque facades and mostly on money-laundering. This monstrous project, deeply rooted in corruption and nationalism truly happened. So, it can happen anywhere. [*With sharp movement, she removes that piece of paper and another blank piece of paper appears underneath. As she writes, she reads*] Act 3. The Yard. [*The installation slowly moves backwards*]

THE PERFORMER: [*Leaning on the moving installation*] Just like this building, the memory moves in a non-linear narrative. As Godard said. . . [*With sharp movement, she points to the audience the cardboard on which 'Everything has a beginning, a middle and an end, but not necessarily in that order' is written — in the front, on the left and on the right, for 5 seconds to each side. She puts down the cardboard*] Godard's *Breathless*, the movie that marked the new wave in the French cinema as a revolutionary example of nonlinear film narration and use of the Jump-Cut approach in editing was one of the two most often shown movies in the cinema of the Railway Residential Building during the 1960s. The other one was Fasbinder's *Love is colder than death*. [*She flips and rises the cardboard. On the other side is the poster for the movie* Love is colder than death. *With sharp movement, she points it to the audience in the front, on the left and on the right, for 5 seconds to each side. She throws the cardboard on the floor. She welcomes the audience to get inside*] You can come closer if you want. Please, take few seconds and try to imagine a building in the center of the cities that you come from that has its own green courtyard. [*Pause. The Performer takes out the cardboard and a hole appears in the installation. She reaches inside and takes out a pot with a bonsai tree and holds it in her hands*] Just as you imagined, this space is used by the tenants for gathering.

Fig. 4: Act 3: Yard. Photo by Nemanja Knezević.

In the 1960s they even parked their cars here. But since of course, there were not many people in Skopje who had cars in the 1960s, that was not a problem. Kovachev Deco from the entrance no. 7, used this as a business opportunity and dug a canal in the yard to repair the broken cars of his neighbors. In the yard, tenants gathered to play cards, chess, to make ajvar and rakia. Nada from the entrance no. 5, then a sports teacher in the neighboring elementary school Pestaloci, here in the yard, taught the tenants to play baseball. In the spring of 1968, at each corner of the yard, one linden was planted. During the 1960s, 400 children were born in the building. It is said that on the occasion of the birth of his first son, Deko buried a bottle of rakia somewhere in the yard. In October 2018, considering that a more than 50-year-old rakia is probably really good, the Macedonian actresses Jasmina Vasileva, Sanja Arsovska, Ivana Pavlaković, Dolores Popović, Tamara Ristovska and Blagoj Veselinov decided to try to find the bottle and dig it out.

A video from the site specific performance If Buildings Could Talk *performed in the Railway Residential Building in November 2018 is projected on the floor. In this video one of the actors quotes a famous scene from Bertoluci`s* Novecento, *where Donald Sutherland and officer Attila is explaining to few nicely dressed gentlemen how should the society finally solve the issue of Communism. After about a minute, Samuel Barber's Adagio fades in [it interferes with the adagio scene on video from the site specific performance]. The Performer comes out, stops on the spot where video is projected, reaches out and throws a lamb of soil on the ground.*

THE PERFORMER: The Railway Residential Building survived the catastrophic earthquake that happened in Skopje in 1963. [*One by one, she throws black cardboards with white numbers on the floor*] 7, 14, 39, 79 tenants didn't survive. The lindens were cut for safety reasons. [*A woman comes out with a tray with rakia and sweets and serves the audience*] The 80% destroyed city was rebuilt with the help from 78 countries from around the world and become the capital of the Yugoslav and World Solidarity. [*The stage-workers start to move the installation to the front and then flip it for 180°, the wall/installation is positioned horizontally. The Performer gets inside. Pause. With a sharp movement, she rises a cardboard on which* Act 4 Terraces *is written*] Act 4! Terraces! [*She holds it for 5 seconds and then puts it down*]

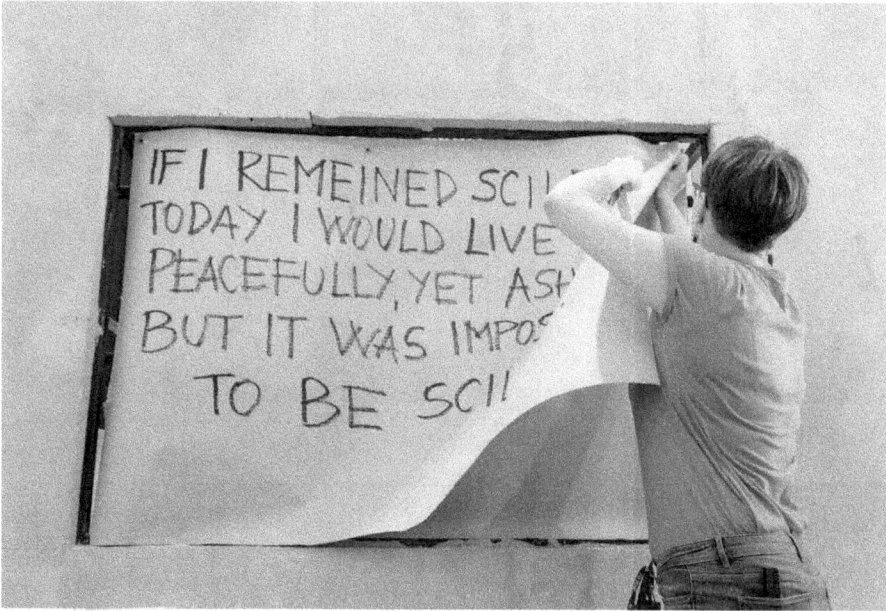

Fig. 5: Act 4: Terraces. Photo by Nemanja Knezević.

THE PERFORMER: You can come closer to watch from every side. We will see each other since these are the open terraces of the Railway Residential Building. Some tenants still remember clearly how they ran out on the terraces one early morning in 1974, when suddenly 'The Internationale' echoed loudly across the yard. They spend hours trying to find the loudspeaker from which it was played. After 2 hours and 17 minutes, finally Schultz found the loudspeaker on one of the upper terraces, but he couldn't stop the music. So he decided to shot a total of 8 bullets in it. After the eight one, the music finally stopped. It was at the end of that summer, when the tenants learned that this scene was fixed by Antonio Kuzmanovski's grandson, just one night after he watched Fellini's *Amarcord* in the building's cinema. You know the movie for sure. You remember the scene, right? Few fascists are hysterically running around a small Italian town, trying to find a loudspeaker from which 'The Internationale' is played. And you must remember the song, right? [*She puts on the worker's topcoat, takes the microphone and sings 'The Internationale'. The stage-workers start to move the installation down (vertical direction). She finishes the song and stands still while the installation is finally set down. Then she takes of the topcoat and puts down*

the microphone. With sharp movement, she rises a cardboard on which 'Act 5. Cinema' is written]
Act 5! Cinema! [*She holds it for 5 seconds and then throws the cardboard on the floor*]

Fig. 6: Act 5: Cinema. Photo by Nemanja Knezević.

THE PERFORMER: Please take few seconds and try to imagine a building in the center of the cities that you come from that has its own cinema. [*Pause*] You've imagined it. This space of more than 150 m2 is used by the tenants today to hold their household council meetings. But they used to organize different events here, gatherings, dance parties, even one wedding was organized here at the Railway Residential Building's cinema.

It is said that in tho 70s, there was a tenant, some remember hiWWm as Kiro, some as Slobodan who kept 49 cats in his apartment. That is strange. But stranger is that in the winter of 1976, tenants frequently found killed cats in the courtyard, and often saw Kiro or Slobodan walking around with a bloody forehead. Nobody knew what was going on until Professor Nada and Blagica, Bajo's wife, met this Kiro or Slobodan on the screening of Bertolucci's *Novecento* at the Cinema. They say that he got particularly excited as he watched the scene [*she takes out a cardboard with a picture of Donald Sutherland in Bertolucci's* Novecento] in which the fascist officer Attila is explaining to few nicely dressed gentlemen how should the society finally solve the issue of Communism. He compares the Communists to tiny pussycats, so he takes one tiny pussycat, hangs it on a wall, runs to it, and kills it with his forehead. [*With a sharp movement, she puts down the cardboard on the installation. From the other side there is a picture of Robert De Niro in* Novecento]

In the mid-1980s, a scene from an important movie for the history of Macedonian cinema was shot here in the cinema of the Railway Residential Building. It is the movie *Happy New Year 1949* — a drama of two brothers of different orientation and fate, which takes place in an atmosphere of tension and fear, during the conflict of Yugoslav Communist Party with the Stalinism. The film is directed by the most prominent Macedonian film director and pedagogue Stole Popov, the son of Trajče Popov that I told you to remember in the start. [*A stage-worker is taking out a small screen on which the scene is played*] We prepared the scene so you can see it. Bajo, Ivan and Tome were among the 32 tenants show up as extras in this scene. If you come closer and listen carefully, you can hear the lovely song. [*Pause. The audience is watching the scene. The Performer takes the microphone and starts to sing the same song* Piši mi *as performed by the Yugoslavian pop singer Nena Ivošević in1980. While she sings, the stage-workers are putting screens on the installation, on which videos used or developed in the different phases of the project* If Buildings Could Talk *are shown. Then, she puts down the microphone, and with a sharp movement she rises a cardboard on which 'Act 6. Building is Community' is written*] Act 6! Building as Community! [*She holds it for 5 seconds and then throws the cardboard on the floor*]

Fig. 6: Act 6: Building as Community. Photo by Nemanja Knezević.

THE PERFORMER: [*She puts away the screen with the scene from* Happy New Year 1949 *and puts in the front another one on which a video from the site-specific performance* If Buildings Could Talk *is shown*] On 1 October, 2017 the site-specific performance *If Buildings Could Talk* had its premiere in the Railway Residential Building. It is a performative tour in which the audience is guided through the many layers of the history about this building. When performed it always ends here in the cinema. The performance was part of a project developed in close collaboration with the tenants of the Railway Residential building, it is inspired by the building, dedicated to the building, has the building as its central character and in a main role. The Railway Residential Building in our performance is what the cherry orchard is in the Chekhov's *Cherry Orchard*. We actors do not act in this performance, we plead, we do not speak, we advocate — for the importance of keeping the public spaces open, against the rigid agendas of the mighty corporations and the corrupt governments, for the political power of communities and their collective actions. In a country that tends to rewrite history, to destroy records, to erase memories, to forget, we are trying to preserve the collective memory into a multi-layered archive. [*Pause. She speaks on the microphone*] The building is utopia. The building is our time. The building can talk. The building talks truly. The building is ideology. The building is politics. The building is policy. The building is falling apart. The building is forgotten by the authorities. The building is ruined by its tenants. The building belongs to those who care for it. The building is the people. The building is the state. The building is culture. The building is art. The building is architecture. The building is public space. The building is ideology. The building is friendship. The building is solidarity. The building is community. [*With a sharp movement, she takes out a small cardboard on which* IF WE CAN DO IT HERE, WE CAN DO IT EVERYWHERE *is written*] IF WE DID IT HERE, WE CAN DO IT EVERYWHERE! [*She holds it for 5 seconds and then, with a sharp moment puts it down on the installation. She puts on the topcoat, quickly bows to the audience and disappears*]

The End.

Note: Based on the concept of Filip Jovanovski, the script was written by Filip Jovanovski and Kristina Lelovac, having Ivana Vaseva and Biljana Tanurovska Kjulavkovski as consultants on the content and Vlado Karaev as a collaborator on the performance script. The installation is a result of the collaboration with the editor Gorjan Atanasov and the sound designer Marko Naumovski. The participants in the production of the pavilion are: Toni Trajkovski (Constructions design dooel, Skopje), Vlado Dimoski, Ivancho Velkov, Andzela Petrovska, Membraning Studio Skopje. The project *This Building Talks Truly* won the prestigious Golden Triga prize for best exposition at Prague Quadrennial of Performance Design and Space 2019.

Karaev, Ilija Tiricovski, Dejan Ivanovski, Sasko Poter Micevski, tenants from the Railway Residential Building

(Slobodan Kocvikj, Tome Karevski, Ivan Dzijanovski) and Oliver Musovikj.

COMING FROM DIRT, WORKING WITH DIRT, WRITING THE DIRT [EXTRACTS, ARIES SEASON, 2019]

AHHAAHHA

minor thoughts on gentrifications and ex-pressions by a multiply-minoritized unbelonger based in E1, Whitechapel, London, UK or an attempt at free-writing and integrating or unfiltrations

I'm writing. I'm writing and...

Speak your heart. Don't bite your tongue. Don't get it twisted. Don't misuse it.

[...]

Am I supposed to change? Are you supposed to change?

Who should be hurt? Who should be blamed?

Am I supposed to change? Are you supposed to change?

Who should be hurt? Who should be ashamed?

Am I supposed to change? Are you supposed to change?

Who should be hurt? Will we remain?

You need a resolution, I need a resolution,

We need a resolution, We have so much confusion. [1]

As I write this, with Aaliyah playing in the background, hoping for a sleep-deprived, generally-deprived burst of energy to spur a stream of writing that can be worked with, that can be edited, I was thinking maybe I should change the music to something more instrumental, electronic, with a beat and rhythm that can catalyze the writing process and get the words down. Then I actually (re)tune into the lyrics being sung by Aaliyah, and realized that they were speaking to exactly what I wanted to write about (tho in a quite different context). Kinda pleading, kinda assertive, kinda up for taking responsibility, kinda wanting some accountability. Kinda it's all too much tbh, kinda I wanna sort this shit out. Questioning herself, questioning who is meant to be hurt, who is allowed to be hurt, whose hurt always gets prioritized. Why? Who always has to be malleable, to contort, who gets to stay static wherever the fuck they are. Over the

---

1   Aaliyah, 'We Need a Resolution', *Aaliyah*. [CD] Los Angeles, CA: Blackground/Virgin, 2001.

non-listening, but still resolved to resolve. Communication channels blocked thru passive aggressions, maintained thru passive aggressions. So appropriate. It makes me think about deeper listening. How easy it is to ignore what is being said right in front of you in favor of your potentially misfired aims and prioritized desires. How sometimes what you want is actually in front of you and the onus is on the viewer/seeker/listener to tune in better to the frequency around them, instead of trying to adapt their surroundings to their own frequencies. I think about how this relates to the gentrification happening in my area, how this is one of issues — people refusing to tune in to where they are. Wanting to reset the dial. Set the agenda. Engage the ego and the savior complex. Save the streets by keeping them clean of dirt, of unnecessary.

W ith its bantering stallholders loudly extolling the virtues of their fruit and veg and excitable teenage girls sifting through piles of knock-off designer bags, Roman Road Market is not a place to go to nurse a hangover on a Saturday morning.

It might be one of the last surviving pockets of authentic East End London, resisting the gentrification that has embraced much of Tower Hamlets, but Bethnal Green's resistance is weakening, wooed by a series of major new developments and fashionable small conversions that are warming it up for a place in the league table of hotspots.

You have only to look at the location, three miles north-east of Charing Cross, and its proximity to Shoreditch and Brick Lane, to know that no matter how shabby the shopping in Bethnal Green Road, regardless of how gritty and untrendy its streets, where you'll search in vain for a chain coffee shop or artisanal bakery, it is still a place that is ready to rock.

Fig. 1: An extract from an article from the Homes and Properties section of the Evening Standard newspaper based in London.[2]

I'm writing. And as I write I resist that internal desire to write academically, to write in a way that what I have to say will be seen as more valid by those I am normally in company with in spaces like this — conferences, summits, workshops. I worry about the overly personalized subjective tone of the piece, worry about having my contribution diminished because of its subjective, anecdotal nature. That it will be seen as less valid knowledge. I worry about this despite knowing that no one has the knowledge I do because no one has lived the life that I have lived (self-evident), but moreover that my knowledge that comes from my lived experience living in social housing in an area experiencing extreme gentrification, my experience working with young people in multiple areas in London going through these shifts and ethnic cleansings, that this knowledge is valid knowledge. I know there is a disdain within academia for the overly

2   Ruth Bloomfirld, 'Hotspot in Waiting: Bethnal Green set to rival Shorditch with trendy bars and new homes in the Victorian chest hospital', *Evening Standard*, 10 April 2018, https://www.standard.co.uk/ homesandproperty/buying-mortgages/bethnal-green-set-to-rival-shoreditch-with-trendy-bars-and-new-homes-in-the-victorian-chest-hospital-a119346.html.

subjective account, especially when it comes from people of color, women, minoritized folk in general. I also understand how the heightened focus on the 'I' among minoritized writers presents issues also. I know it isn't seen as real knowledge. It confuses me. Cos there's still an impulse in me that thinks that if I back this up with references and citations, I will be more credible. As if the work of PhD students and academics, especially when it comes to working-class communities of color, has more validity than my voice, just because it's evidenced or peer-reviewed or whatever. When I know that most middle-class people actually have no idea how to talk to working-class people. When I know that councils around the country lie about the extent of consultation they do with working-class young people. (They do nothing. Is basics the answer). When deprived communities are apparently being offered a platform to express their voice, but are actually just being talked on behalf of, by the paternalistic, white, middle-class charity sector. What's even real? In the context of 'what's even real?', why am I still doubting my realness and elevating academia to realness? The levels. The bullshit. Imposter syndrome down to the bone.

Speak your heart. Don't bite your tongue. Don't get it twisted. Don't twist your tongue, don't bite your heart to mould yourself around these people. These people who speak confidently on behalf of you but have no idea who you are. Don't do it! Be real be real be real. Am I supposed to change or are YOU supposed to change? Who should be hurt? Who should be blamed? Who are the actual criminals? Who is the danger? Will we remain? I know we're already being made to disappear in so many ways. Will we remain? I dunno I dunno...

Sajid Javid cuts funding for knife crime programme

Caroline Wheeler, Deputy Political Editor

Sunday February 03 2019, 12.01am GMT, The Sunday Times

Fig. 2: A headline from an article on The Times website.[3]

We have so much confusion.

When you give a talk now about these kinda themes — working with young people, encouraging creativity among working-class, Black and Brown youth —

you always look at the news to see if there's a relevant news story. There always is. For this panel (which is almost every panel), I mentioned a story that appeared on Resident Advisor.

---

3   Caroline Wheeler, 'Sajid Javid Cuts Funding for Knife Crime Programme', 3 February 2019, https://www. thetimes.co.uk/article/sajid-javid-cuts-funding-for-knife-crime-programme-nx7vpzkh3.

YouTube removes over 30 UK drill
rap videos that police say incite
violence

Fig. 3: A headline from an article on Resident Advisor.[4]

When creative expression and criminality become blurred, is it safe to speak? The met police claim these videos incite violence and therefore should be censored. I don't see the white-man-heart-free-speech-brigade now. You think of the youth you work with, ones who make music videos on their estates, who write raps and poems and choreograph, who speak about their experiences of their ends or just having fun. How important this is to them. How accessible this form of creativity is for them. And then how this, like everything else related to Black and Brown youth, especially young men, just becomes criminalized. Every day, every day. You think of what it means to have had a childhood in the ends, and a teenagehood in white elite society. How you have never seen more criminality than in the basements of Oxbridge rented accommodation. How middle/upper-class white boys can literally be found guilty of dealing drugs, sometimes sent to prison, then get to return to finish their degree and continue their middle-class white trajectory. Every day, every day. How middle-upper-class white boys can literally write a dramatic "fictionalized" public account of their illegal drug dealing, and recount how they got away with it because they are middle-upper-class white boys, and again still get to go back to Oxbridge university and finish their degrees. It's not fucking fair that people in my area die for the same things that just allow these white boys to live/thrive. Which bodies get criminalized and punished? Which don't? Not the game but the undesirable players. The exploitable players. Straight up never seen as many drugs in my life as I did in Oxbridge. White powders up white nostrils. White nostrils that become white lawyers and white bankers and white filmmakers and white artists and white academics and white facilitators and white charity sectors.

Brown boys falling off balconies. Neighbors ask if it was ganja or phagul. Weed or crazy. Both? More?

You go to a performance in a gallery in Bethnal green featuring queer/trans Black people and People of Color speaking out about the violence they face as a result of how they (choose to) express themselves. The clothes that I maybe incorrectly see as adornment rather than reality cos I don't have the space to understand where I'm at yet and this constricts my vision. The mental struggle of understanding the struggle of being urself when ur self gets violence just for being itself. This pain you feel 2 deeply on the inside. Internally. On the way home I notice a bunch of police officers talking to two Black boys off Globe Road. I don't wanna be the one to intervene but everyone else is just walking past. People always walk past. Too many times

---

4   Andrew Ryce, 'YouTube removes over 30 UK drill rap videos that police say incite violence', 30 May
       2018, https://ra.co/news/41855.

in recent memory of intervening before police violence. Cba to engage with them. They claim they are doing important work in ridding the area of knives and drugs, doing stop and search on the basis of intelligence. People of color are always adorned with something that will invite violence regardless of how many layers you take off or how conforming your clothes is. The endless violence of the Skin U r in. Externally. I stick around till they're gone. Try asking the boys if they're cool. They almost shrug it off. 'Every day, every day'.

You go to a inter faith meeting where a representative from the Met is talking about hate crimes in ur gentrified ends. How the group the Met is specifically and especially focusing on atm is the LGBT community. There's been an increase in hate crime against this group. The numbers of religious hate crimes and racist hate crimes (these two things are treated as separate categories — you wonder how?) were far far higher esp. since Brexit. You remember the Peter Tatchell walk in ur ends.

You ask the youth at a community center in East to rate their area in terms of safety. They all give it 9 or 10 out of 10. You enquire further and they explain all the things they're scared of. Real shit. But that this is just their normal. Their life. They see it as nothing extra-ordinary.

You apply to become a teaching assistant in Germany. It's a lottery deciding which area you get assigned to. You end up in Dresden, home of PEGIDA and birthplace of organized neo-Nazi and islamophobic sentiment. You are 100% certain no other Muslim people applied for this program. You express your fears about being sent to a place like that to the White German woman who organizes this, and why specifically you with your visible Brownness and therefore assumed Muslimness got put into that locality. Your disquiet gets routinely erased, and your combined identities invite the laughter of disbelief, of 'r u kidding me? ur all of that?' She claims they can do nothing for you, until at some point you re-iterate your minoritized gender. At which point a flurry of expression and possibility: 'Oh yes of course, maybe we can do something for you now, because this fits into our diversity objectives.' So Muslim and brown is appaz not diversity but T is?? How some parts of you are seemingly so accommodated whilst other parts of u are thrown out the house. Life and death eating each other inside U. Is that life? WTF is home? Where do u go?

Summer 2017 and hijabi family afraid to go outside. Acid attacks. Stuck in their domestic space. Where do u go?

You're walking home on Friday evening down Bethnal Green road. You see rainbow flags outside what seems like pretty much every single establishment. You hear loud club music emanating from random establishments along the way - kebaberies cum discotheque. You are bombarded with signs of who is being welcomed in your area and also who is disappearing. Where do u go? Where are you meant to go? You dunno. Make home make place in the schism between. Eroding shoreline. Where do the working class QTIPOC go? The ones in between.

You wake up in mid-afternoon and hear voices from behind the curtain. Two Bengali babies informing each other about what to do in case of an acid attack. 'Stay on that side of the pavement.' 'Keep away from car windows'. 2k17 childhood.

Everyone I mean every single person leaving Bethnal Green station looks like a hipster every single time u exit the station. Every single time. Where do u go?

You wanted to write WHITE before hipster above and still hesitated/censored yourself. Why?

You wanna dance, you wanna dance, you wanna try move your body. You realize that almost every space of liberation is a space of gentrification. Where do u go?

Your friend tells you about this new megaclub in East. How security herd white techno lovers into the safety of the inside, keep them away from the dangerous unnecessary dirt, whilst the Met harass these Black and Brown boys existing in public, stop and search. Where do u go?

Something weird about the levels of danger and unsafety and violence that these shifts you see signify. When things are getting cosmetic. When people like the new shops. When you know what it means. Underneath. But you don't know what it means. How you see your community under so much attack that you cut your own throat out to try speak for them, with them, protect them. What's going on?

You see posters in the street advertising the new inhabitants. You don't know if you ever knew what irony meant but you don't even wanna engage with that now. You see commissioned graffiti talking about sex with refugees. You think you should be going crazy. Mouth agape like WTF is actually going on?

Fig. 4: A billboard in Whitechapel, East London advertising new office containers. Photo by AHHAAHHA.

Fig 5: A row of people waiting under a commissioned work of public 'art' on a black wall with the words 'SEX WITH REFUGEES IS JASMINE-SCENTED AND BEAUTIFUL'

No space anywhere. Free from violence.

You keep wondering, how can art and academia actually serve these spaces? What can it do for these spaces? With these spaces? Is it part of the merry go round? Mind melts.

This is the context we exist in. That artists and academics and facilitators and charity sectors operate in. This is the levels of endless violence young working-class kids exist in. This doesn't even come close to explaining all the levels. This is what you need to understand if you're gonna work with them i.e. not save them. You talk to other friends also working with/trying to engage young men of color. They all reluctantly whisper about how impossible it is. How no one is really engaging with that impossibility. How hopeless it is.

I made a shift last November of no longer talking to audiences (esp. more academic audiences) about good practice in working with young, Black, Brown, working-class communities and instead choosing to focus on the need to deconstruct and shift patterns of thinking — get out of Whiteness. Otherwise the endless merry go round of White Saviors and gatekeepers keeps on jingling. Who is supposed to change?

Was reading today about Avatar — apparently there is going to be a sequel to the White Savior Fantasy. On and on and on and on.

I tried on some level to engage with some of the violences I experience, rather than that of the youth I work with, and realized that I was/still am one of those young people engaging with society and white charity sector violencenonsense doing traumatizing diversity youth work. That I don't need to talk for other people as a facilitator, when I can talk about myself as a youth participant, as a young creative, as someone they're gonna work with/for/on, someone that also experiences on different levels the excesses of violence that becomes ordinary and mundane for poor folk. After an incident in November 2018 involving crazy white violence directed at me that I CBA to get into fully RN, I was done. Done done done. [Briefly put, a white couple verbally abusing me and my disability, and later calling the police three times primarily because they got so racially stressed out by being described as white. WHITE.] Done done done. Over the merry-go-round.

I want people to appreciate that violence. Appreciate those barriers to expression. Understand how free some people's speech always is.

You think about all the white saviors you've become acquainted with. Why it's easier for white artists to work with refugees and those that won't speak back, but will never engage with the kids in ends. How malleable our tongues become. How unwelcome our whispers. Contortion artists. How gratitude is one of the powerfullest weapons in the hands of the have-it-alls. Leaving us to beg for scraps and curtsy whilst doing so.

So often at these panels, you know you're in the POC panel. You know you are an afterthought, even in these most radical academic spaces. You think what hope is there? Are you always already marginalized? Or is this a bate example of active marginalization by people who cannot tune in to our frequency. Always on the margins. So often the only POC in the room. That weight. Those hungry eyes. They wanna devour my experience. I can feel it. I know it. It is exhausting. Cos it's everything and everywhere. You know you shouldn't be critical of these spaces that have invited you in. Scratch that. You know you should be critical of these spaces that have marginalized you in, but you know it's not the proper thing to be critical of these spaces that think they have graciously invited you in. Sticks in the throat. After years and years of a too short freelance career of being the only POC on the panel, or being on the POC afterthought panel of the white conference, you know you gotta speak summit akin to ur truth...

'Can I talk to you? Comfort you?'

You know that there's no point of being on this endless merry-go-round. The circle that goes round and round. That surreptitious sigh. The darting eyes to check and to avoid. A resigned chuckle. An endless consumption of resources. This surely isn't the revolution. Can we plz resolve to make sure the merry go round doesn't become the revolution? Feel good. Re-trauma. Does it even belong to me anymore? It's in the marketplace now. In the door, consume the poverty, go back to settled lives. That meme about the big secret in academia. Get to feel good, whilst we just continue dying. A friend framed the conversation around gentrification as slow, deliberate murder. This shifted it. I see the informal community centers. Elders and the betis from numbers 1, 3, 7 perched around the Bengali Channel TV or quranic reading judgements or Muslim advice or random documentaries of village life. A strange double alienation. Estranged. I see how these moments are what sustains life for people who can't access other kinds of state services. For people actively discriminated against in housing, health, education. How when people are made to disperse, they lose those spaces of connection. Those things that sustain life. Slow death. Murder.

White gentrifiers next door complain about the noise from the TV - this informal community center for Bangladeshi neighbours. Is it the noise, or is it the unintelligibility? The life? They don't even speak to us. Complain straight to the council. Bedrock of instability. A waft of skunk smoke wades in from their balcony. Every night. You can do illegal if you're white.

Don't wanna contribute to this anymore. Don't wanna just package it all again. For the vultures. They wanna devour. When everyone is a vulture. Says it's saving you as it pecks out your entrails. NAH NAH NAH. TBH I'm bored. Bored of trying to package complexities of experiences, that I tbh only know through my specific and also somewhat alienated lens, and tell people how to work with marginalized youth. Working-class youth.

Exhausted.

They just go back to their homes. They get to feel good. They get to consume. In peace. In security.

Ain't that the same as the european gentrifiers next door? Get to go back to their homes. Feel good. Feel good in ways that send black and brown boys to the cell or down the drain. Brain cells gone. Consume. In peace. In security. Everything is being adapted for them.

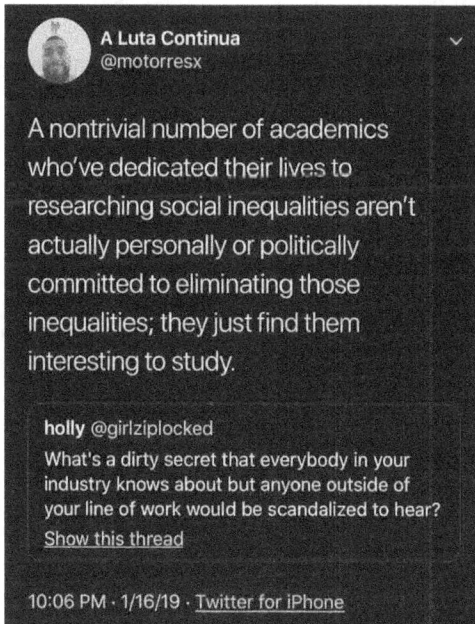

A Luta Continua
@motorresx

A nontrivial number of academics who've dedicated their lives to researching social inequalities aren't actually personally or politically committed to eliminating those inequalities; they just find them interesting to study.

holly @girlziplocked
What's a dirty secret that everybody in your industry knows about but anyone outside of your line of work would be scandalized to hear?
Show this thread

10:06 PM · 1/16/19 · Twitter for iPhone

Fig.6: A tweet by @motorresx on the lack of commitment to ending inequalities by academics who research them.

Life can be so solid for some. So stable. Even tho it's all resting on this bedrock of violence and instability. Somehow they don't feel it. Like actually WTF. Nothing to say but WTF.

There's a fire in ur estate. 5am and the whole block evacuated. Freezing outside. You realize this is the first time you've seen this new make up of residents, everyone out of the boxes. You take some time to notice the different colors, the different languages. How odd to be forced to actually engage in visual contact with the inhabitants of the boxes next to, above and below you. Your amma and other neighbors huddle in some auntie's house in the block opposite, rounds of tea and biscuits to warm worried, cold limbs.

Mobility. You think about the car as informal community center for brown boys who wanna smoke and got nowhere else to go. The TBQH is that the car does better outreach and space holding than any community center — at least when it comes to brown boys...

'Every day every day'

You're at an urban planning conference taking place in a site of gentrification. There's less than 10 POC. You kinda hate the fact that you always count but you're always gonna. Archive this sh*t. A bunch of white people talking about planning the future. This hasn't ever gone wrong. At one point someone talks about the meta levels of a performance of white people consulting the public i.e. the 99% white public about an urban plan. How people were forced to consider the performativity of consultation, how this performance reflects the reality. No one seemed to get or express on how many other levels this entire performance, not just the specific performance but the entire conference performance, reflects reality — a bunch of white people talking to other white people planning shit for everyone speaking on behalf of everyone, pretending that everyone is in the room or not even noticing the bate absence of melanated folk. I'm not gonna be the person to say it. Do they notice us when we are not there?

DONE DONE DONE.

Too poor to play: children in social housing blocked from communal playground

Fig. 7: A headline from the Guardian website on inequalities in social housing.[5]

---

5   Hariet Grant and Chris Michael, 'Too Poor to Play: Children in Social Housing Blocked from Communal Playground' *The Guardian* 25 March 2019, https://www.theguardian.com/cities/2019/mar/25/too-poor-to-play-children-in-social-housing-blocked-from-communal-playground.

You're at a cafe in Hackney to meet with a white, male friend of yours who studied architecture and was working for a firm based in East/Hackney. He tells you about the firm he's just started working for, how they're 'redeveloping' a community space, how they produced a proposal for the new space with many many images over many many pages of what the space will look like and who will inhabit it. He tells you there was not a single person of color / Black person in the entire publication. How he had to be the one to tell them what area they were actually in. Do they notice us when we are not there?

I DON'T NEED TO DECORATE MY BODY FOR IT TO BE DESTROYED.

I DON'T NEED TO DECORATE MY BODY FOR IT TO BE DESTROYED.

I DON'T NEED TO DECORATE MY BODY FOR IT TO BE DESTROYED.

A strange spectral presence. Hypervisible and monstrously numbered but also invisible as if never there at all. Over-existing and ripe for destruction. Superfluous.

I think about 'blacklisting'. How when you speak out, you get cut off. Made (slowly) dead. How many acquaintances have told me how they have materially lost out because of attempting to check the racism/-ism of an institution or space. How much POC expression is tinged with violence. The threat to cut you off. How I'm worried this might happen here and how I'm thinking about what I'm writing and how I'm trying to speak with truth but how I'm sure I still probably cushioned it for those in need of mollycoddling cos that fear is too real. The actual snowflakes. How white tears burn and ain't worth it. How splitting it is to cut yourself off for fear of being cut off. How difficult it is to engage in creative expression when you twist your tongue for sport. Performance. Nothing will change in this endless spectacular merry go round unless deep shifts in thinking are engaged with. Part of what needs to catalyze that shift is the (im)possibility of unbridled POC expression. Without fear of punishment or retribution from sensitive white ears. No more questioning yourself. No worry if this is a mess. It's ok.

Every day, every day.

You're saying no. You're learning to say no. To refuse. To engage in the language of refusal. You think about how powerful it is to say no, how, just like the word sorry, you were not taught to say this word. In the context of 13+ people in 3 bedroom cramped council flat, boundaries and privacy and the capacity to say NO, to reinforce and recreate boundaries was fantastical, never possible. Membranes. Separations. Refusal. You worry about writing this still. You worry that it's too real, that it's not wanted. In other words, that you're too real, and you're unwanted. What does it look like to say I'm here and I'm real. Even tho you know how powerful it is to refuse. How much it will do for you if you refuse. Don't need to go along with the merry go round. It hasn't served you thus far. Engage your power. Say no!

I DON'T NEED TO DECORATE MY VOICE FOR IT TO BE DESTROYED.

I DON'T NEED TO DECORATE MY VOICE FOR IT TO BE DESTROYED.

I DON'T NEED TO DECORATE MY VOICE FOR IT TO BE DESTROYED.

I'm thinking about freewriting. Freewriting as an emancipatory practice and exercise in practicing unbridled unrestricted speech. Speech from the self not for the other. Speech inhibited by the other in the self but not the other in the other.

I recently started performing free writes. More out of a lack of care and time from white colonial organizations rather than a self-directed and desired mode of performance. It feels uncomfortable. Too raw. Too unedited.

But I'm thinking now, how this is a helpful mode of communication/production for someone who has been crippled by white colonial academic institutions and rendered unable to write. To produce. Forcing others to engage with my mess in the same way I had to engage with white mess in academia. Maybe it's not all understandable. Maybe it's messy and unrefined. Maybe that's fine.

Leaving home. Making new homes.

Scrambling to understand how u r meant to engage with integration. Is it possible? Splitting apart. You can't integrate all of this.

You have so much confusion.

Learning what dirt could look like.

Every day, every day.

ACKNOWLEDGEMENTS

I would like to say my thank yous to friends and comrades who consciously or unconsciously influenced my thinking and this book. I would like to thank Elena Marchevska for her selfless support as my supervisor and friend during my Mari Curie fellowship at London South Bank University where this book was conceived and my friends and comrades from Southwark Notes (Rastko Novaković, Chris Jones, Mara Ferreri and Caterina Satori-Khan), who made our stay in an unwelcoming London survivable for us East European migrants. I would also like to thank Anja Buechele and Matthew Hyland for their friendship and readiness to (proof)read my lousy article drafts whenever I asked. I also want to thank my comrades from the Radical Housing Journal for teaching me about housing struggles and how to work together in the last three years. I thank my companions from the Roof—anti-eviction organization in Serbia, especially Vladimir Logos, Saša Perić, Tanja Šljivar, Vladimir Mentus, Nemanja Pantović for their solidarity, and Irena Ristić for reading and commenting on this introduction. I want to thank all those who said yes to my last-minute proofreading and image file formatting invite, especially Liz Mason-Deese, Cam Neufeld, Magid Shihade, Tanja Juričan, Tijana Parezanović, as well as others that wanted to stay anonymous; together with anonymous peer-reviewers they truly made this essay collection into a collective endeavor. My thanks go to numerous editors that rejected this manuscript but pushed me to think further, most of all to Pia Pol, and big thanks to the Institute of Network Cultures for loving this book. A final thank you goes to my son Vedan Šurelov for surviving my experimental approach to mothering, and to my mum Vukica Vilenica, for being there for me.

BIOGRAPHIES

AHHAAHHA is a writer from E1, London. AHHAAHHA does youth and care work, on the 'frontlines,' in the shadows. AHHAAHHA assumes apocalypses. AHHAAHHA trusts in youth, wants to create and destroy worlds with youth (the future). AHHAAHHA is interested in strategies for appearance and disappearance of hurt bodies. AHHAAHHA is interested in tongues untied, in human and non-human becoming.

Josephine Berry is Research Tutor in the School of Arts and Humanities at the Royal College of Arts and Research Lab Lecturer for the MA Cultural Industry at Goldsmiths University in London, focusing on experimental methods of research and a spatial approach to culture. She has worked as an editor for the cultural politics magazine *Mute* since 1995. Her monograph *Art and Bare Life: A Biopolitical Inquiry* (2019) brings the biopolitical theory to bear on aesthetic theories of autonomous art. She is currently working on a book that explores the elusive subject of Planetary Realism.

Alyssa Erspamer lives in Rome where she works as a junior facilitator for the Matter Group. She completed her postgraduate degree at the University of Manchester in Visual Anthropology and her undergraduate degree at University College London in European Social and Political Studies, with a dissertation in Art History. She grew up in Boston, Massachusetts, to Italian parents. Her research interests are in leftist theory/movements, bureaucratic and humanitarian institutions/organizations, religion, Italian traditions, contemporary art/aesthetics, and applied continental philosophy. At the time of writing, she is working on a short film for Poesia in Azione's series, Città a parole.

Ioana Florea has been researching social inequalities in Romanian cities and spatial processes of social differentiation. Since 2006, she has been involved as an activist with several grassroots groups and collective political projects from Bucharest working and struggling for social justice, such as Our City – Our Decision, the Political Art Gazette, and the Common Front for Housing Rights. Ioana is active as a militant researcher with the Block for Housing national platform and the European Action Coalition for the Right to Housing and the City. Since 2017, she has been a Post-doctoral Researcher at the Department of Sociology and Work Science, University of Gothenburg.

Nicola Guy is a writer, curator, and educator based in the UK. She is a PhD History candidate at the University of Hull and a member of the AHRC project The Heritage Consortium. Her thesis 'Art, Activism or Advertising? The role of exhibition-making in unified Berlin' looks at different ways in which contemporary art exhibitions have contributed to building a collective identity in post-1989 Berlin. She has previously held positions at Nottingham Contemporary, Flat Time House, London and continues to work with Archive Books, Berlin, as well as contributing to exhibitions and public programs at different institutions.

Lucas S. Icó is from Rio de Janeiro, Brazil. He currently lives in Porto Alegre. He works as an artist, researcher, designer, and teacher. He holds a master's degree in visual arts from

PPGAV-EBA-UFRJ. His work is based on developing ideas and projects thinking through contemporary aesthetics in different communities and in contexts of political challenge.

Anthony Iles is currently Commissioning Editor for the series Documents of Contemporary Art
published by the Whitechapel Gallery and MIT Press, an Associate Lecturer at Northampton
University and a Visiting Tutor at the School of Art & Design, Middlesex University. He was a
founding member of Full Unemployment Cinema and a Contributing Editor with *Mute / Met-
amute* since 2005, he is the author, with Josephine Berry, of the book, *No Room to Move: Art
and the Regenerate City* (2011), co-author, with Tom Roberts, of *All Knees and Elbows* (2012)
contributing editor to the anthologies, *Anguish Language: Writing and Crisis* (2015), and *Look
at Hazards, Look at Losses* (2017). Anthony recently completed a PhD entitled *Paper Assembly* on the artist-run journal *Inventory* (1995-2005) in 2019.

Filip Jovanovski is a visual artist and architect from Bitola/Skopje. He graduated at the Faculty
of Architecture in Skopje and gained his MA thesis at the Faculty of Fine Arts in Skopje. He
was one of the authors of the Macedonian Pavilion titled *Freeing Space* which was presented
at the Venice Biennale for architecture in 2018. He has made about 20 stage designs for
theatre plays, video and documentary projects and won several awards. Since 2007, he is
the artistic director of the AKTO Festival for contemporary arts in Bitola and is co-managing
the organization for art and culture 'Faculty of things that cannot be learned (FR~U).'

Sylvi Kretzschmar has been investigating the use of PA systems to amplify the human voice,
particularly in political forums and demonstrations. Her artistic work with the MEGAPHONE
CHOIR is a facet of her practice around the PA in political contexts, where choreography
consistently merges with a political movement. During 2012-2015 she was a scholarship
holder of the artistic-academic postgraduate program 'Assemblies and Participation: Urban
Publics and Performance'. Kretzschmars' work bridges the fields of electronic music, live art,
and choreography. Together with Camilla Milena Fehér, she is SKILLS, an artistic duo who
create music through physical action, movement, and dance.

Kristina Lelovac is an actress and assistant professor of acting at the Faculty of Dramatic
Arts in Skopje. She is enrolled in doctoral studies in Theatrology. Her fields of interest are
professional training of actors and rethinking of theatre practices in the context of (contemporary) political realities. She has participated in domestic and international festivals, study
visits, summer schools, and workshops (ARENA Internationales Festival für Tanz, Theater
und Performance - Erlangen, IMPULSE - Düsseldorf, US State Department IVL Program). She
is a member of several activist initiatives and one of the founders of the Festival of Feminist
Culture and Action 'Firstborn Girl.'

Andreea S. Micu is a Post-doctoral Fellow in the Mahindra Humanities Center at Harvard
University. Her research examines the intersection of aesthetics, migration, and activism in
the south of Europe in the aftermath of the 2008 economic crisis. Her book manuscript, *The
Performative Commons: Housing Activism and Aesthetics in the Austere City*, documents case
studies in which working-class groups ally across ethnic differences to imagine alternatives
to neoliberal urban development and (re)build the urban commons in Madrid, Rome, and

Athens. Micu's work has appeared in venues such as *Performance Philosophy* and *pARTic-ipatory Urbanisms*. She holds a PhD in Performance Studies from Northwestern University.

Veda Popovici is a political worker based in Bucharest. She is a member of various anti-author-itarian, anarchist, and feminist collectives such as Macaz cooperative and collective, Dysno-mia collective and the Political Art Gazette and has taught classes on decolonial thought, nationalism, and feminist theory at the National University of Arts in Bucharest and the Uni-versity of California Santa Cruz. Dedicated to radical housing action, she has cofounded the Common Front for Housing Rights in Bucharest, the Block for Housing (a national anti-cap-italist platform) and currently is the facilitator of the European Action Coalition for the Right to Housing and the City.

Cristina T. Ribas works as an artist and researcher and often writes. She holds a PhD in Art from Goldsmiths College University of London (2017) and MA from the Instituto de Artes / UERJ (2008). She has been working with cartography, feminism, and militant research related to artistic production and has been learning from improvisation theatre to develop other forms of non-discursive research. In 2011, Cristina created the open platform Desarquivo.org. From 2017 she engaged in Arquivos Táticos with Giseli Vasconcelos and Tatiana Wells. She is also part of the network Red Conceptualismos del Sur and from Associação Imotirô.

Ivana Vaseva is a curator and researcher of cross-disciplinary, collaborative, and socially engaged works and programs from Skopje. She is the program director of the organization Faculty of things that cannot be learned (FR~U).' Vaseva is co-curator of AKTO Festival for contemporary arts existing since 2006. She won the award *Ladislav Barishikj* of AICA – Mace-donia for the research project 'Collective actions as a political, and not organizational decision' (2015, co-author) and Special Architecture Award from the Association of Architects of Mace-donia (2014, co-participant). Vaseva also specialized in the Curatorial Program (2011/2012) at de Appel Arts Centre in Amsterdam, the Netherlands.

Ana Vilenica is an urban and cultural researcher with a research interest in grassroots cul-tural and political action against capitalist urban and housing regimes. Vilenica is the Radical Housing Journal Editor and the Editor for Central and East Europe at Interface-journal for and about social movements. She edited four books, among which *Becoming a Mother in Neoliberal Capitalism* (uz)bu))na))), 2013, 2016), *On the Ruins of Creative City* (kuda.org, 2013) and *Fragments for Studies on Art Organizations* (kuda.org, 2020). She is a long-term housing and feminist activist in Serbia, the UK, and on the European level. Most recently, she has been organizing with the Roof anti-eviction organization from Serbia, EAST – Essen-tial Autonomous Struggles Transnational and the European Action Coalition for the Right to Housing and the City.

School of Echoes Los Angeles is a multi-racial and cross-generational autonomous collective of organizers, teachers, and sometimes artists, who formed in 2012 around community-based research and organizing against gentrification in Los Angeles. In 2015, the School of Echoes founded the LA Tenants Union, a diverse movement of tenants across the city of Los Angeles building tenant power through advocacy, education, and direct action.

Southwark Notes are a small network of researchers and writers who have been active in the North Southwark area of South London since 2010, exposing and actively protesting the social cleansing of the area. They post research on their popular blog southwarknotes.wordpress.com which also functions as a repository of archival material of local urban struggles as well as features a detailed look at the complicit role of art in the service of *regeneration* and displacement.

You Should See the Other Guy are an all womxn, queer led theatre collective led by Nina Scott and Emer Morris: multidisciplinary artists and housing campaigners working with London communities affected by displacement, gentrification, and social cleansing. Since 2015, they have been delivering embedded community theatre projects which work on the ground and on the stage to tackle housing injustice. They are community organizers as well as writers, directors, Theatre of the Oppressed practitioners, song makers, poets, set designers, and human rights practitioners. Their work aims to demonstrate that powerful, potent, and beautiful theatre belongs in community-led spaces, not only the elite spaces of traditional theatre.

www.ingramcontent.com/pod-product-compliance
Lightning Source LLC
Chambersburg PA
CBHW062051270326
41931CB00013B/3031